Lecture Notes in Computer Science 10373

Commenced Publication in 1973
Founding and Former Series Editors:
Gerhard Goos, Juris Hartmanis, and Jan van Leeuwen

Juan de Lara · Detlef Plump (Eds.)

Graph Transformation

10th International Conference, ICGT 2017
Held as Part of STAF 2017
Marburg, Germany, July 18–19, 2017
Proceedings

 Springer

Editors
Juan de Lara 🆔
Universidad Autónoma de Madrid
Madrid
Spain

Detlef Plump 🆔
University of York
York
UK

ISSN 0302-9743   ISSN 1611-3349   (electronic)
Lecture Notes in Computer Science
ISBN 978-3-319-61469-4   ISBN 978-3-319-61470-0   (eBook)
DOI 10.1007/978-3-319-61470-0

Library of Congress Control Number: 2017944215

LNCS Sublibrary: SL1 – Theoretical Computer Science and General Issues

Printed on acid-free paper

This Springer imprint is published by Springer Nature
The registered company is Springer International Publishing AG
The registered company address is: Gewerbestrasse 11, 6330 Cham, Switzerland

Foreword

Software Technologies: Applications and Foundations (STAF) is a federation of leading conferences on software technologies. It provides a loose umbrella organization with a Steering Committee that ensures continuity. The STAF federated event takes place annually. The participating conferences may vary from year to year, but they all focus on foundational and practical advances in software technology. The conferences address all aspects of software technology, from object-oriented design, testing, mathematical approaches to modeling and verification, transformation, model-driven engineering, aspect-oriented techniques, and tools.

STAF 2017 took place in Marburg, Germany, during July 17–21, 2017, and hosted the four conferences ECMFA 2017, ICGT 2017, ICMT 2017, and TAP 2017, the transformation tool contest TTC 2017, six workshops, a doctoral symposium, and a projects showcase event. STAF 2017 featured four internationally renowned keynote speakers, and welcomed participants from around the world.

The STAF 2017 Organizing Committee would like to thank (a) all participants for submitting to and attending the event, (b) the Program Committees and Steering Committees of all the individual conferences and satellite events for their hard work, (c) the keynote speakers for their thoughtful, insightful, and inspiring talks, and (d) the Philipps-Universität, the city of Marburg, and all sponsors for their support. A special thanks goes to Christoph Bockisch (local chair), Barbara Dinklage, and the rest of the members of the Department of Mathematics and Computer Science of the Philipps-Universität, coping with all the foreseen and unforeseen work to prepare a memorable event.

July 2017   Gabriele Taentzer

Preface

This volume contains the proceedings of ICGT 2017, the 10th International Conference on Graph Transformation. The conference was held in Marburg, Germany, during July 18–19, 2017. ICGT 2017 was affiliated with STAF (Software Technologies: Applications and Foundations), a federation of leading conferences on software technologies. ICGT 2017 took place under the auspices of the European Association of Theoretical Computer Science (EATCS), the European Association of Software Science and Technology (EASST), and the IFIP Working Group 1.3, Foundations of Systems Specification.

The aim of the ICGT series is to bring together researchers from different areas interested in all aspects of graph transformation. Graph structures are used almost everywhere when representing or modelling data and systems, not only in computer science, but also in the natural sciences and in engineering. Graph transformation and graph grammars are the fundamental modelling paradigms for describing, formalizing, and analyzing graphs that change over time when modelling, e.g., dynamic data structures, systems, or models. The conference series promotes the cross-fertilizing exchange of novel ideas, new results, and experiences in this context among researchers and students from different communities.

ICGT 2017 continued the series of conferences previously held in Barcelona (Spain) in 2002, Rome (Italy) in 2004, Natal (Brazil) in 2006, Leicester (UK) in 2008, Enschede (The Netherlands) in 2010, Bremen (Germany) in 2012, York (UK) in 2014, L'Aquila (Italy) in 2015, and Vienna (Austria) in 2016 following a series of six International Workshops on Graph Grammars and Their Application to Computer Science from 1978 to 1998 in Europe and in the USA.

This year, the conference solicited research papers describing new unpublished contributions in the theory and applications of graph transformation, innovative case studies describing the use of graph transformation techniques in any application domain, and tool presentation papers that demonstrate the main features and functionalities of graph-based tools. All papers were reviewed thoroughly by at least three Program Committee members and additional reviewers. We received 23 submissions, and the Program Committee selected 14 papers for publication in these proceedings, after careful reviewing and extensive discussions. The topics of the accepted papers range over a wide spectrum, including theoretical approaches to graph transformation and their verification, model-driven engineering, chemical reactions, as well as various applications. In addition to these paper presentations, the conference program included an invited talk, given by Georg Gottlob (University of Oxford, UK).

We would like to thank all who contributed to the success of ICGT 2017, the invited speaker, Georg Gottlob, the authors of all submitted papers, as well as the members of the Program Committee and the additional reviewers for their valuable contributions to the selection process. We are grateful to Reiko Heckel, the chair of the Steering Committee of ICGT for his valuable suggestions; to Javier Troya and Leen Lambers for their help in preparing the proceedings; to Gabriele Taentzer, the general chair of STAF; and to the STAF federation of conferences for hosting ICGT 2017. We would also like to thank EasyChair for providing support for the review process.

July 2017   Juan de Lara
                                                                          Detlef Plump

Organization

Steering Committee

Michel Bauderon	LaBRI, University of Bordeaux, France
Paolo Bottoni	Sapienza University of Rome, Italy
Andrea Corradini	University of Pisa, Italy
Gregor Engels	University of Paderborn, Germany
Holger Giese	Hasso Plattner Institut Potsdam, Germany
Reiko Heckel (Chair)	University of Leicester, UK
Dirk Janssens	University of Antwerp, Belgium
Barbara König	University of Duisburg-Essen, Germany
Hans-Jörg Kreowski	University of Bremen, Germany
Ugo Montanari	University of Pisa, Italy
Mohamed Mosbah	LaBRI, University of Bordeaux, France
Manfred Nagl	RWTH Aachen, Germany
Fernando Orejas	Technical University of Catalonia, Spain
Francesco Parisi-Presicce	Sapienza University of Rome, Italy
John Pfaltz	University of Virginia, Charlottesville, USA
Detlef Plump	University of York, UK
Arend Rensink	University of Twente, The Netherlands
Leila Ribeiro	University Federal do Rio Grande do Sul, Brazil
Grzegorz Rozenberg	University of Leiden, The Netherlands
Andy Schürr	Technical University of Darmstadt, Germany
Gabriele Taentzer	University of Marburg, Germany
Bernhard Westfechtel	University of Bayreuth, Germany

Program Committee

Anthony Anjorin	University of Paderborn, Germany
Paolo Baldan	University of Padova, Italy
Gábor Bergmann	Budapest University of Technology and Economics, Hungary
Paolo Bottoni	Sapienza University of Rome, Italy
Andrea Corradini	University of Pisa, Italy
Juan de Lara (Co-chair)	Autonomous University of Madrid, Spain
Juergen Dingel	Queen's University, Canada
Rachid Echahed	CNRS, Laboratoire LIG, France
Maribel Fernandez	King's College London, UK
Holger Giese	Hasso Plattner Institut Potsdam, Germany
Joel Greenyer	Leibniz University of Hannover, Germany
Annegret Habel	University of Oldenburg, Germany

Reiko Heckel	University of Leicester, UK
Berthold Hoffmann	University of Bremen, Germany
Dirk Janssens	University of Antwerp, Belgium
Barbara König	University of Duisburg-Essen, Germany
Leen Lambers	Hasso Plattner Institut Potsdam, Germany
Yngve Lamo	Bergen University College, Norway
Mark Minas	University of Bundeswehr München, Germany
Mohamed Mosbah	LaBRI, University of Bordeaux, France
Fernando Orejas	Technical University of Catalonia, Spain
Francesco Parisi-Presicce	Sapienza University of Rome, Italy
Detlef Plump (Co-chair)	University of York, UK
Arend Rensink	University of Twente, The Netherlands
Leila Ribeiro	University Federal do Rio Grande do Sul, Brazil
Andy Schürr	Technical University of Darmstadt, Germany
Uwe Wolter	University of Bergen, Norway
Albert Zündorf	University of Kassel, Germany

Additional Reviewers

Azzi, Guilherme	Maximova, Maria
Debreceni, Csaba	Nolte, Dennis
Dyck, Johannes	Peuser, Christoph
Farkas, Rebeka	Rabbi, Fazle
Flick, Nils Erik	Raesch, Simon-Lennert
Gadducci, Fabio	Sandmann, Christian
Gritzner, Daniel	Schneider, Sven
Heindel, Tobias	Szárnyas, Gábor
Kluge, Roland	

General and Fractional Hypertree Decompositions: Hard and Easy Cases (Invited Talk)

Wolfgang Fischl[1], Georg Gottlob[1,2], and Reinhard Pichler[1]

[1] Institut für Informationssysteme, Technische Universität Wien, Vienna, Austria
[2] Department of Computer Science, University of Oxford, Oxford, UK

Abstract. Hypertree decompositions [5, 4], the more powerful generalized hypertree decompositions (GHDs) [5, 4], and the yet more general fractional hypertree decompositions (FHDs) [8, 9] are hypergraph decomposition methods successfully used for answering conjunctive queries and for solving constraint satisfaction problems. For a survey on these decompositions, see [3], for exact hypertree decomposition algorithms [7], and for heuristic decomposition methods, see [1].

Each hypergraph H has a width relative to each of these methods: its hypertree width $hw(H)$, its generalized hypertree width $ghw(H)$, and its fractional hypertree width $fhw(H)$, respectively. While $hw(H) \leq k$ can be checked in polynomial time, the complexity of checking whether $fhw(H) \leq k$ holds for a fixed constant k was unknown. We settle this problem by proving that checking whether $fhw(H) \leq k$ is NP-complete, even for $k = 2$ and by same construction also the problem deciding whether $ghw(H) \leq k$ is NP-complete for $k \geq 2$. Hardness was previously known for $k \geq 3$ [6], whilst the case $k = 2$ has remained open since 2001.

Given these hardness results, we investigate meaningful restrictions, for which checking for bounded ghw is easy. We study classes of hypergraphs that enjoy the bounded edge-intersection property (BIP) and the more general bounded multi-edge intersection property (BMIP). For such classes, for each constant k, checking whether $ghw(H) \leq k$, and if so, computing a GHD of width k of H is tractable and actually FPT. Finally we derive some approximability results for fhw. We consider classes of hypergraphs whose fhw is bounded by a constant k and which also enjoy the BIP or MIP, or bounded VC-dimension. For each hypergraph in such a class, we are able to compute an FHD of width $O(k \log k)$ efficiently. A different restriction on classes of hypergraphs gives a linear approximation in PTIME. Hypergraphs of bounded rank are a simple example of such a class.

A full paper [2] with these and further results is available online via the link https://arxiv.org/abs/1611.01090.

The work of Fischl and Pichler was supported by the Austrian Science Fund (FWF):P25518-N23. Gottlob's work was supported by the EPSRC Programme Grant EP/M025268/ VADA: Value Added Data Systems Principles and Architecture

References

1. Dermaku, A., Ganzow, T., Gottlob, G., McMahan, B., Musliu, N., Samer, M.: Heuristic methods for hypertree decomposition. In: Proceedings of MICAI, pp. 1–11 (2008)
2. Fischl, W., Gottlob, G., Pichler, R.: General and fractional hypertree decompositions: hard and easy cases. CoRR, abs/1611.01090 (2016)
3. Gottlob, G., Greco, G., Leone, N., Scarcello, F.: Hypertree decompositions: questions and answers. In: Milo, T., Tan, W. (eds.) Proceedings of the 35th ACM SIGMOD-SIGACT-SIGAI Symposium on Principles of Database Systems, PODS 2016, San Francisco, CA, USA, 26 June–01 July 2016, pp. 57–74. ACM (2016)
4. Gottlob, G., Leone, N., Scarcello, F.: A comparison of structural CSP decomposition methods. Artif. Intell. **124**(2), 243–282 (2000)
5. Gottlob, G., Leone, N., Scarcello, F.: Hypertree decompositions and tractable queries. J. Comput. Syst. Sci. **64**(3), 579–627 (2002)
6. Gottlob, G., Miklós, Z., Schwentick, T.: Generalized hypertree decompositions: Np-hardness and tractable variants. J. ACM **56**(6), 1–32 (2009)
7. Gottlob, G., Samer, M.: A backtracking-based algorithm for hypertree decomposition. ACM J. Exp. Algorithm. **13** (2008)
8. Grohe, M., Marx, D.: Constraint solving via fractional edge covers. ACM Trans. Algorithms **11**(1), 1–20 (2014)
9. Marx, D.: Approximating fractional hypertree width. ACM Trans. Algorithms, **6**(2), 1–17 (2010)

Contents

Model Transformation and Tools

Foundations

The Pullback-Pushout Approach to Algebraic Graph Transformation

Andrea Corradini[1]($\boxtimes$), Dominque Duval[2]($\boxtimes$), Rachid Echahed[2]($\boxtimes$),
Frédéric Prost[2]($\boxtimes$), and Leila Ribeiro[3]($\boxtimes$)

[1] Dipartimento di Informatica, Università di Pisa, Pisa, Italy
andrea@di.unipi.it
[2] CNRS and Université Grenoble Alpes, Grenoble, France
{Dominique.Duval,Rachid.Echahed,Frederic.Prost}@imag.fr
[3] INF - Universidade Federal do Rio Grande do Sul, Porto Alegre, Brazil
leila@inf.ufrgs.br

Abstract. Some recent algebraic approaches to graph transformation include a pullback construction involving the match, that allows one to specify the cloning of items of the host graph. We pursue further this trend by proposing the Pullback-Pushout (PB-PO) Approach, where we combine smoothly the classical modifications to a host graph specified by a rule (a span of graph morphisms) with the cloning of structures specified by another rule. The approach is shown to be a conservative extension of AGREE (and thus of the SQPO approach), and we show that it can be extended with standard techniques to attributed graphs. We discuss conditions to ensure a form of locality of transformations, and conditions to ensure that the attribution of transformed graphs is total.

1 Introduction

Algebraic graph transformations have been dominated by two main approaches, namely the Double Pushout (DPO) [9] and the Single Pushout (SPO) [14]. These two approaches offer a very simple and abstract definition of a large class of graph transformation systems [5,8]. However, they are not suited for modeling transformations where certain items of the host graph should be copied (cloned), possibly together with the connections to the surrounding context. This feature is instead naturally available in approaches to graph transformation based on node replacement, like Node-Label-Controlled (NLC) grammars [10], and is needed in several application domains. The NLC approach is typically presented in set-theoretical terms, but a categorical formulation was proposed in [1]. The key points there are that a rule is represented as a morphism from the right-hand side (rhs) to the left-hand side (lhs) (both enriched to represent abstractly the possible embedding context), and a match is a morphism *from* the host graph to

This work has been partially supported by the LabEx PERSYVAL-Lab (ANR-11-LABX-0025-01) funded by the French program Investissement d'avenir and by the Brazilian agency CNPq.

J. de Lara and D. Plump (Eds.): ICGT 2017, LNCS 10373, pp. 3–19, 2017.
DOI: 10.1007/978-3-319-61470-0_1

the lhs. Then rewriting is modeled by a pullback: the cloning of edges due to node replacement is obtained by the multiplicative effect of the limit construction.

Independently, some approaches were proposed to enrich DPO with cloning, including Adaptive Star Grammars [6], Sesqui-Pushout (SQPO) [4] and AGREE [3]. Even if the presentations differ, all are based on the idea of introducing a limit construction in the first phase of rewriting, to model cloning. Coherently with the DPO, in these approaches a match is a morphism from the lhs of the rule *to* the host graph but, at least for SQPO and AGREE, the match determines implicitly a morphism *from* the host graph to an enriched version of the lhs, which is pulled back along a suitable morphism to model deletion and cloning of items. Other approaches to structure transformations where the match goes *from* the host graph to the rule include [19] for refactoring object-oriented systems, and [20] for ontologies: in both cases some form of cloning can be modeled easily.

The analysis of these approaches led us to define (yet) an(other) algebraic approach to graph transformation, called PB-PO, that we introduce in this paper. The PB-PO approach conservatively extends AGREE [3], and thus SQPO [4] with injective matches, by streamlining the definition of transformation and making explicit the fact that when cloning is a concern, it is natural to include in the transformation a pullback construction based on (part of) the match, that has to go *from* the host graph to (part of) the lhs of the rule. A rule in PB-PO is made of two spans, the *top* and the *bottom* ones, forming two commutative squares. A match consists of a pair of morphisms, the *instantiation* from the lhs of the top span to the host graph (like a standard match in DPO and similar approaches), and the *typing* from the host graph to the lhs of the bottom span, that is used to clone items with the first phase of a transformation, which is a pullback. As the name of the approach suggests, the second phase is a standard pushout which glues the pullback object with the rhs of the top span. Thus a PB-PO transformation can be seen as a combination of a standard transformation of structures, modeled by the top span, with a sort of retyping modeled by the bottom span.

Like other categorical approaches supporting cloning (e.g. [1,3]) also PB-PO may specify transformations that are not local, in the sense that they affect part of the host graph that is not in the image of the instantiation. After showing in which sense the new approach extends AGREE, we propose a formal notion of locality for PB-PO rules and a sufficient condition to ensure it.

Next we consider the enrichment of the PB-PO with attributes, following the ideas developed in [7] for the SQPO approach. A key feature of this approach is to allow attributes of items of the host graph to be changed through the application of a rule, a feature that is possible thanks to the use of *partially* attributed structures. As a consequence, in general the result of transforming a completely attributed graph via a PB-PO rule could be a partially attributed graph. We present some sufficient syntactic conditions over rules in order to ensure that the result of a transformation is totally attributed.

The paper is organized as follows: In Sect. 2, we define PB-PO rewriting and in Sect. 3, we show its relation with the AGREE and SQPO approaches. Then, we discuss issues regarding the locality of PB-PO rewriting in Sect. 4. In Sect. 5, we show how the PB-PO approach extends to deal with attributed structures. Finally, we conclude in Sect. 6.

2 The PB-PO Transformation of Structures

In this section we introduce the PB-PO approach to structure transformation. The main differences with respect to other algebraic approaches is the shape of a rule and, as a consequence, the definition of a match. To make the presentation lighter, we start assuming that all objects and diagrams belong to a category of "structures" $\mathbf{G}$ with "enough" pullbacks and pushouts so that the required constructions exist. We will introduce any additional requirement on $\mathbf{G}$ when needed. Typical examples of categories of interest are that of graphs, of hyper-graphs, or of typed graphs (i.e., the slice category $\mathbf{Graph} \downarrow T$ for a given type graph T). Such categories have all limits and colimits.

Definition 1 (Rule, Match and Rewrite Step). *A PB-PO rule ρ is a commutative diagram as follows:*

$$
\begin{array}{ccccc}
L & \xleftarrow{\quad l \quad} & K & \xrightarrow{\quad r \quad} & R \\
\downarrow{\scriptstyle t_L} & = & \downarrow{\scriptstyle t_K} & = & \downarrow{\scriptstyle t_R} \\
L' & \xleftarrow{\quad l' \quad} & K' & \xrightarrow{\quad r' \quad} & R'
\end{array}
\tag{1}
$$

We call $L \xleftarrow{l} K \xrightarrow{r} R$ the top span *of ρ and $L' \xleftarrow{l'} K' \xrightarrow{r'} R'$ its* bottom span. *The three vertical arrows are called the* left-hand (lhs) side, *the* interface *and the* right-hand (rhs) side *of ρ. We say that ρ is in* canonical form *if the left square is a pullback and the right square is a pushout.*

A (PB-PO) match of ρ in an object G is a factorization of its left-hand side through G, i.e. a pair (m, m') such that $m' \circ m = t_L$, as shown on the right. Arrow $m : L \to G$ is called the instantiation *(part) and arrow $m' : G \to L'$ the* typing *(part) of the match.*

$$
t_L \left(= \begin{array}{c} L \\ \downarrow{\scriptstyle m} \\ G \\ \downarrow{\scriptstyle m'} \\ L' \end{array} \right.
$$

A PB-PO rewrite step from G to H via rule ρ, denoted $G \Rightarrow_\rho H$, is defined by the following diagram, where square (a) is a pullback, arrow $n : K \to D$ (making square (a') commuting) is uniquely determined by the universal property of pullbacks, square (b) is a pushout, and arrow $p' : H \to R$ makes square (b') commuting and is uniquely determined by the properties of pushouts.

$$
\begin{array}{ccccc}
L & \xleftarrow{\quad l \quad} & K & \xrightarrow{\quad r \quad} & R \\
\;\;\downarrow{\scriptstyle m} & =(a') & \downarrow{\scriptstyle n} & PO\ (b) & \downarrow{\scriptstyle p} \\
G & \xleftarrow{\quad g \quad} & D & \xrightarrow{\quad h \quad} & H \\
\;\;\downarrow{\scriptstyle m'} & PB\ (a) & \downarrow{\scriptstyle n'} & =(b') & \downarrow{\scriptstyle p'} \\
L' & \xleftarrow{\quad l' \quad} & K' & \xrightarrow{\quad r' \quad} & R'
\end{array}
\qquad (2)
$$

Note that if rule ρ is in canonical form and it is applied to a match (m, m'), in the resulting Diagram (2) we have that (a') is a pullback and (b') is a pushout by obvious properties of these universal constructions.

It is worth observing that object R' of a rule is not involved directly in a rewrite step, but it determines a default typing for the result of rewriting. Thus a PB-PO rewrite step maps a PB-PO match (m, m') to another PB-PO match (p, p').

Example 2. This example is related to the copy of some "local" web pages, as discussed in [3]. Nodes represent web pages, and edges represent hyperlinks among them. Then it is reasonable to expect that creating a copy of a set of local pages will only copy the hyperlinks contained in such pages, and not those in remote pages pointing to them. This is modeled by the following PB-PO rule ρ in the category of graphs, where the vertical morphisms map $\underline{r}, \underline{r}', \underline{n}$ respectively to $\underline{l}, \underline{l}', \underline{n}$. Note that in order to avoid confusion between morphism names and graph node names, the latter will be underlined in the rest of the paper.

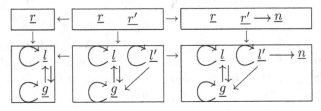

A match $L \xrightarrow{m} G \xrightarrow{m'} L'$ classifies the nodes of G as either *local* ($\underline{l}$) or *global* ($\underline{g}$) thanks to the typing $m' : G \to L'$ and it distinguishes one *root* node ($\underline{r}$) of G thanks to the instantiation $m : L \to G$. In addition, $\underline{r}$ is local since $m' \circ m = t_L$. The *local subgraph* of G is defined as the subgraph of G generated by the local nodes. By applying the rule ρ to the match (m, m') we get a graph H which contains G together with a copy of its local subgraph with all its outgoing edges, and with an additional edge from the copy of the root to a new node $\underline{n}$. Here is an instance of such a rewrite step, where the root of G is $\underline{r}_1$, its local nodes are $\underline{r}_1, \underline{l}_1, \underline{l}_2$ and its global nodes are $\underline{g}_1, \underline{g}_2$:

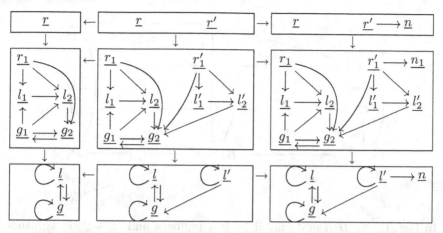

A natural question is whether rules in canonical forms are as expressive as general rules. The following result answers positively to this question.

Proposition 3 (canonical forms). *For each* PB-PO *rule ρ there exists a rule ρ_1 in canonical form which is equivalent, that is*

- *ρ and ρ_1 have the same lhs $t_L : L \to L'$*
- *for each match (m, m') with $L \xrightarrow{m} G \xrightarrow{m'} L' = t_L$, $G \Rightarrow_\rho H$ if and only if $G \Rightarrow_{\rho_1} H$.*

Proof. The following diagram shows how one can build from rule ρ (whose components are named as in Diagram (1)) a corresponding rule ρ_1 (where corresponding components have subscript 1).

$$
\begin{array}{ccccc}
L & \xleftarrow{\quad l \quad} & K & \xrightarrow{\quad r \quad} & R \\
\downarrow{\scriptstyle id_L} & = & \downarrow{\scriptstyle n} & PO & \downarrow{\scriptstyle p} \\
L_1 = L & \xleftarrow{\quad l_1 \quad} & K_1 & \xrightarrow{\quad r_1 \quad} & R_1 \\
\downarrow{\scriptstyle t_{L_1}=t_L} & PB & \downarrow{\scriptstyle t_{K_1}} & PO & \downarrow{\scriptstyle t_{R_1}} \\
L_1' = L' & \xleftarrow{\quad l_1'=l' \quad} & K_1' = K' & \xrightarrow{\quad r_1' \quad} & R_1' \\
& & & r' & \searrow \\
& & & & R'
\end{array}
$$

First rule ρ is applied using the PB-PO approach to match (id_L, t_L), generating the pullback object K_1 and the pushout object R_1. Next we build the pushout of r_1 and t_{K_1}, obtaining object R_1'. It is obvious by construction that rule ρ_1, made of top span $L_1 \xleftarrow{l_1} K_1 \xrightarrow{r_1} R_1$ and of bottom span $L_1' \xleftarrow{l_1'} K_1' \xrightarrow{r_1'} R_1'$, is canonical, and also that the lhs of ρ and ρ_1 coincide.

Now let G be an object and (m, m') be a PB-PO match of ρ (and ρ_1) in G, and consider the following diagram. We argue that $G \Rightarrow_\rho H$ if and only if $G \Rightarrow_{\rho_1} H$.

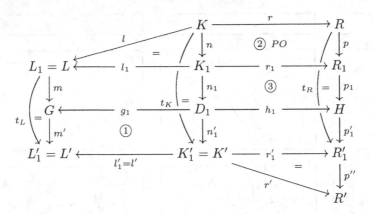

In fact, $G \Rightarrow_\rho H$ if and only if ① is a pullback and ② + ③ is a pushout, while $G \Rightarrow_{\rho_1} H$ if and only if ① is a pullback and ③ is a pushout. Thus we can conclude by observing that since ② is a pushout by construction, ③ is a pushout if and only if ② + ③ is a pushout, by well known properties of composition and decomposition of pushouts.   □

Example 4. It is easy to check that the following three PB-PO rules are equivalent and act as identities on any graph with at least one node. The last one is in canonical form.

$$\boxed{\underline{n}} \leftarrow \boxed{} \rightarrow \boxed{} \qquad \boxed{\underline{n}} \leftarrow \boxed{\begin{matrix}\underline{n}\\ \underline{n'}\end{matrix}} \rightarrow \boxed{\begin{matrix}\underline{n}\\ \underline{n'}\end{matrix}} \qquad \boxed{\underline{n}} \leftarrow \boxed{\underline{n}} \rightarrow \boxed{\underline{n}}$$

$$\boxed{\underline{n}\ \underline{g}} \leftarrow \boxed{\underline{n}\ \underline{g}} \rightarrow \boxed{\underline{n}\ \underline{g}} \qquad \boxed{\underline{n}\ \underline{g}} \leftarrow \boxed{\underline{n=n'}\ \underline{g}} \rightarrow \boxed{\underline{n=n'}\ \underline{g}} \qquad \boxed{\underline{n}\ \underline{g}} \leftarrow \boxed{\underline{n}\ \underline{g}} \rightarrow \boxed{\underline{n}\ \underline{g}}$$

3   Relating PB-PO with AGREE and SQPO Rewriting

In this section, we first show that PB-PO extends both AGREE and SQPO with monic matches (in categories where AGREE rewriting is defined), and then discuss informally how the greater expressive power can be exploited in designing transformation rules.

An AGREE rule α [3] is a triple of arrows with the same source, as in the left part of (3), and its application to an AGREE match $m : L \rightarrowtail G$ is shown in the right part of (3), defining a rewrite step $G \Rightarrow_\alpha^{\text{AGREE}} H$ as explained below.

$$L \xleftarrow{\ l\ } K \xrightarrow{\ r\ } R \qquad\qquad L \xleftarrow{\qquad l \qquad} K \xrightarrow{\qquad r\qquad} R \qquad (3)$$

Thus an AGREE rule is made of the usual top span enriched with a *mono* $t : K \rightarrowtail T_K$ having a role similar to arrow $t_K : K \rightarrow K'$ in a PB-PO rule. The definition of rewriting requires the existence in the underlying category of a *partial map classifier* [2], i.e. for each object Y, there exists an arrow $\eta_Y : Y \rightarrowtail T(Y)$ such that for each pair of arrows $Z \xleftarrow{i} X \xrightarrow{f} Y$ (a *partial map* from Z to Y) there is a unique arrow $\varphi(i,f)$ such that the left diagram of (4) is a pullback.

$$
\begin{array}{ccc}
X \xrightarrow{\ f\ } Y & \quad L \xleftarrow{\ l\ } K & \quad L \xrightarrow{\ id_L\ } L \qquad (4)\\
\scriptstyle i\downarrow \quad \lrcorner \quad \downarrow\scriptstyle \eta_Y & \scriptstyle \eta_L\downarrow \qquad \downarrow\scriptstyle t & \scriptstyle m\downarrow \quad \lrcorner \quad \downarrow\scriptstyle \eta_L\\
Z \xrightarrow[\varphi(i,f)]{} T(Y) & T(L) \xleftarrow[l'=\varphi(t,l)]{} T_K & G \xrightarrow[\overline{m}=\varphi(m,id_L)]{} T(L)
\end{array}
$$

The application of the AGREE rule α to a match $m : L \rightarrow G$ is obtained by first taking the pullback (a) of $\overline{m}$ and l', and then the pushout of the resulting mediating arrow n and of r. Both $\overline{m}$ and l' are uniquely determined by $T(L)$ as shown in the mid and right diagrams of (4). By comparing Diagrams (2) and (3) we easily obtain the following result.

Proposition 5 (relating AGREE and PB-PO). *Let α be an AGREE rule in a category with a partial map classifier. Then there is a PB-PO rule ρ_α such that for each mono $m : L \rightarrowtail G$ we have $G \Rightarrow_\alpha^{\mathrm{AGREE}} H$ if and only if $G \Rightarrow_{\rho_\alpha} H$ using match $(m, \overline{m})$ with $\overline{m} : G \rightarrow T(L)$.*

Proof. Let $\alpha = (L \xleftarrow{l} K \xrightarrow{r} R, K \xrightarrow{t} T_K)$ be an AGREE rule, and $T_K \xrightarrow{r'} R' \xleftarrow{t_R} R$ be the pushout of $T_K \xleftarrow{t} K \xrightarrow{r} R$. Let ρ_α be the PB-PO rule having $L \xleftarrow{l} K \xrightarrow{r} R$ as top span, $T(L) \xleftarrow{l'} T_K \xrightarrow{r'} R'$ as bottom span, and η_L, t and t_R as the three vertical arrows relating them. Then the statement immediately follows by comparing Diagrams (2) and (3) defining $G \Rightarrow_{\rho_\alpha} H$ and $G \Rightarrow_\alpha^{\mathrm{AGREE}} H$, respectively.   □

We easily obtain a similar result for SQPO rewriting with monic matches. An SQPO rule has the shape $\sigma = (L \xleftarrow{l} K \xrightarrow{r} R)$ and its application to a match $L \rightarrow G$ is defined in [4] via a double-square diagram where the right square is a pushout (as in DPO), but the left square is a *final pullback complement*. In [3] it was shown that for a monic match $L \rightarrowtail G$, $G \Rightarrow_\sigma^{\mathrm{SQPO}} H$ if and only if $G \Rightarrow_{\alpha_\sigma}^{\mathrm{AGREE}} H$, where $\alpha_\sigma = (\sigma, \eta_K : K \rightarrowtail T(K))$ is obtained by enriching σ with the partial map classifier applied to K. The following result is then obvious.

Corollary 6 (relating SQPO and PB-PO). *Let σ be a SQPO rule in a category with a partial map classifier. Then there is a PB-PO rule ρ_σ such that for each mono $m : L \rightarrowtail G$ we have $G \Rightarrow_\sigma^{\mathrm{SQPO}} H$ if and only if $G \Rightarrow_{\rho_\sigma} H$ using match $(m, \overline{m})$ with $\overline{m} : G \rightarrow T(L)$.*

There is therefore a progressively increasing expressive power moving from DPO to SQPO to AGREE to PB-PO, at least for injective matches. A detailed analysis of the expressive power of PB-PO is a topic of future work, but we make a few considerations with respect to the kind of cloning typical of approaches based on node replacement. Standard DPO with left injective rules cannot model cloning of nodes at all. Instead SQPO can, with a non-injective lhs $l : K \to L$. Referring to the right diagram in (3), if a node $n \in L$ is cloned (i.e., it has more than one inverse image in K via l), then its image in G will be cloned as well in D. Furthermore, for any *embedding edge* e, i.e. an edge incident to $m(n)$ in G but not in $m(L)$, there will be one copy of e in D for each counter-image of $m(n)$.

With AGREE the same kind of node cloning can be specified, but thanks to the additional arrow $t : K \rightarrowtail T_K$ in the rule, one can specify explicitly which embedding edges have to be copied for each cloned node of G. Moving to PB-PO, note that arrows $t_L : L \to L'$ and $l' : K' \to L'$ are explicitly provided by a PB-PO rule, while the corresponding arrows $\eta_L : L \rightarrowtail T(L)$ and $l' : T_K \to T(L)$ in (3) are uniquely determined by $l : K \to L$ and $t : K \to T_K$ in AGREE. With suitable definitions of object L' and arrow $l' : K' \to L'$, and using the m' part of a match, in PB-PO one can

- classify in a fine way the *context items* of the host graph G, i.e. those not in the image of m;
- for each group of such items, specify if it is deleted, preserved or copied;
- specify additional application conditions.

Example 7. Suppose that in an information system there are two security levels: $\top$, for private information, and $\perp$ for public information. We can model the transformation of a graph containing both private and public information nodes so that in the resulting graph there is no access (arrow) from public to private ones. This can be done with a rule having the empty graph for L, K and R, the following inclusion $K' \subseteq L'$ for l', and the identity for r':

$$\boxed{\overset{\curvearrowright}{C}\top \overset{\rightleftharpoons}{\underset{\kappa}{}} \perp \overset{\curvearrowleft}{\supset}} \xleftarrow{\quad l' \quad} \boxed{\overset{\curvearrowright}{C}\top \overset{\curvearrowright}{} \perp \overset{\curvearrowleft}{\supset}}$$

Given a morphism $m' : G \to L'$, mapping all private information nodes to $\top$ and public nodes to $\perp$, the application of this rule to the match $(\emptyset \to G, m')$ would erase all arrows from public nodes to private nodes.

4   Constraining the Effects of PB-PO Rewriting

As just discussed (and evident from Example 7), a PB-PO rewrite step can affect any item of the host graph, that is, changes are not limited to the image of L and its incident edges (as in other approaches like DPO, SPO and SQPO). This holds for AGREE as well, as discussed in [3] where a notion of *local rule* was introduced. Informally, let us denote with $A \setminus B$ the largest subobject of A disjoint from B. Then an AGREE rule is local if for all matches $m : L \to G$ we have that $G \setminus m(L)$

is preserved after the transformation, i.e. referring to Diagram (3), if $D \setminus n(K) \to G \setminus m(L)$ is an isomorphism. Also, in [3] a sufficient condition for an AGREE rule to be local was identified.

In the case of PB-PO, the greater flexibility in the definition of rules and of matches on the one hand allows us to introduce a more general notion of locality, called Γ-*preservation*, parametrized by a subobject $\Gamma \rightarrowtail L'$ of the lhs of the bottom span. On the other hand, however, whether a rewrite step is Γ-preserving or not depends not only on the rule but also on the match. After introducing the notion of Γ-preservation and characterizing a sufficient condition to ensure it, we relate it to locality of AGREE transformations.

Definition 8 (Γ-preserving rewrite steps). *Let ρ be a PB-PO rule as in Diagram (1), inc : $\Gamma \rightarrowtail L'$ be a mono, and (m, m') be a match of ρ in G. Let G_Γ be defined by the pullback on the left of Diagram (5).*

Then we say that the rewrite step $G \Rightarrow_\rho H$ is Γ-preserving if, referring to Diagram (2), the two squares on the right of Diagram (5) are pullbacks.

$$
\begin{array}{ccc}
G_\Gamma \rightarrowtail i \to G & \quad & G_\Gamma \leftarrow id \dashv G_\Gamma \rightarrowtail id \to G_\Gamma \\
\downarrow \quad \lrcorner \quad \downarrow m' & & \downarrow i \quad \downarrow j \quad \downarrow h \circ j \\
\Gamma \rightarrowtail inc \to L' & & G \leftarrow g - D - h \longrightarrow H
\end{array} \qquad (5)
$$

Intuitively, G_Γ represents the subobject of G typed by Γ. The right diagram of (5) says that subobject G_Γ remains unchanged in D (left square) and in the resulting structure H (right square). Note that in Diagram (5) it follows from the three squares being pullbacks that i, j and also $h \circ j$ are mono.

The next result presents sufficient conditions for a rewrite step to be Γ-preserving, under suitable assumptions on the underlying category and on the rule.

Proposition 9 (conditions for Γ-preservation). *Let us assume that the underlying category of structures $\mathbf{G}$ is adhesive [13] and has a strict initial object 0 (i.e., each arrow with target 0 must have 0 as source). Let ρ be a PB-PO rule in canonical form or right-linear, i.e., where r is a mono, and let (m, m') be a match of ρ in G with $m : L \rightarrowtail G$ mono. Then we have that if the two squares of Diagram (6) are pullbacks, then $G \Rightarrow_\rho H$ is a Γ-preserving rewrite step.*

$$
\begin{array}{ccc}
0 \longrightarrow L & \qquad & \Gamma \rightarrowtail id \to \Gamma \\
\downarrow \quad \lrcorner \quad \downarrow m & & \downarrow \quad \lrcorner \quad \downarrow inc \\
G_\Gamma \rightarrowtail i \to G & & K' - l' \to L'
\end{array} \qquad (6)
$$

Informally, the left pullback ensures that the subobject of G typed over Γ is disjoint from the image of L in G, thus it is not affected by the top rule; the right pullback guarantees that the subobject Γ of L' (and thus the items of G typed on it) is preserved identically when pulled back along l', thus it is not affected by the bottom rule. It is still open if the previous result also holds for rules that are neither in canonical form nor right linear. To conclude this section, we relate the PB-PO notion of Γ-preservation with locality of AGREE rules.

Proposition 10 (Γ-preservation and AGREE locality). *Let α be an AGREE rule as in Diagram (3) and ρ_α be the associated PB-PO rule as in Proposition 5. Let Γ_α be the subobject $T(0)$ of $T(L)$. If α is local in the sense of [3], then for each AGREE match $m : L \rightarrowtail G$ the rewrite step $G \Rightarrow_{\rho_\alpha} H$ is Γ_α-preserving.*

5   The PB-PO Transformation of Attributed Structures

For attributed structures we follow the same approach as in [7]. Given a category **G** called the category of structures, with pullbacks and pushouts, a category **A** called the category of attributes, and two functors $S : \mathbf{G} \to \mathbf{Set}$ and $T : \mathbf{A} \to \mathbf{Set}$, the category of attributed structures **AttG** and the category of partially attributed structures **PAttG** are defined as in [7]. The issue is that there are not enough pushouts in the category **PAttG**. Let **Pfn** denote the category of sets with partial maps. A partial map f from X to Y is denoted $f : X \rightharpoonup Y$ and its domain of definition is denoted $\mathcal{D}(f)$. The partial order $\leq$ between partial maps is defined as usual: let $f, g : X \rightharpoonup Y$, then $f \leq g$ means that $\mathcal{D}(f) \subseteq \mathcal{D}(g)$ and $f(x) = g(x)$ for all $x \in \mathcal{D}(f)$. Then **Pfn** with this partial order is a 2-category. By composing S and T with the inclusion of **Set** in **Pfn** we get two functors $S_p : \mathbf{G} \to \mathbf{Pfn}$ and $T_p : \mathbf{A} \to \mathbf{Pfn}$. Let $|...|$ denote any of the four functors S, T, S_p, T_p (sometimes, $|...|$ is omitted).

Definition 11 (attributed structures). *The category of attributed structures **AttG** (with respect to the functors S and T) is the comma category $(S \downarrow T)$. This means that an attributed structure is a triple $\widehat{G} = (G, A, \alpha)$ made of an object G in **G**, an object A in **A** and a map $\alpha : |G| \to |A|$; and a morphism of attributed structures $\widehat{g} : \widehat{G}_1 \to \widehat{G}_2$, where $\widehat{G}_1 = (G_1, A_1, \alpha_1)$ and $\widehat{G}_2 = (G_2, A_2, \alpha_2)$, is a pair $\widehat{g} = (g, a)$ made of a morphism $g : G_1 \to G_2$ in **G** and a morphism $a : A_1 \to A_2$ in **A** such that $\alpha_2 \circ |g| = |a| \circ \alpha_1$. The category of partially attributed structures **PAttG** is defined similarly: a partially attributed structure is a triple $\widehat{G} = (G, A, \alpha)$ made of an object G in **G**, an object A in **A** and a partial map $\alpha : |G| \rightharpoonup |A|$; and a morphism of partially attributed structures $\widehat{g} : \widehat{G}_1 \to \widehat{G}_2$, where $\widehat{G}_1 = (G_1, A_1, \alpha_1)$ and $\widehat{G}_2 = (G_2, A_2, \alpha_2)$, is a pair $\widehat{g} = (g, a)$ made of a morphism $g : G_1 \to G_2$ in **G** and a morphism $a : A_1 \to A_2$ in **A** such that $\alpha_2 \circ |g| \geq |a| \circ \alpha_1$. A morphism of partially attributed structures (g, a) is called strict when $\alpha_2 \circ |g| = |a| \circ \alpha_1$.*

$$
\begin{array}{ccccccc}
\widehat{G}_1 & & G_1 & & |G_1| \xrightarrow{\alpha_1} |A_1| & & A \\
\widehat{g} \downarrow & = & g \downarrow & & |g| \downarrow \quad \geq \quad \downarrow |a| & & \downarrow a \\
\widehat{G}_2 & & G_2 & & |G_2| \xrightarrow{\alpha_2} |A_2| & & A_2
\end{array}
$$

Given an attributed structure $\widehat{G} = (G, A, \alpha)$, we write $\underline{n} : x$ when $\underline{n}$ has attribute x (i.e. $\alpha(\underline{n}) = x$) and $\underline{n} : \perp$ when $\underline{n}$ is not attributed (i.e. $\underline{n} \notin \mathcal{D}(\alpha)$), $|G|_\perp$ denotes the set of elements of $|G|$ which are not attributed. An attributed structure $\widehat{G} = (G, A, \alpha)$ is said attributed *over* A, an attributed morphism $\widehat{g} = (g, a)$ is said attributed *over* a, and when $a = id_A$ then $\widehat{g}$ can be said attributed *over* A.

Remark 12. A morphism of partially attributed structures $\widehat{g} = (g, a) : \widehat{G}_1 \to \widehat{G}_2$ is such that $\widehat{g}(\underline{n}_1 : x_1) = g(\underline{n}_1) : a(x_1)$ and $\widehat{g}(\underline{n}_1 : \perp) = g(\underline{n}_1) : \perp$ or $g(\underline{n}_1) : x_2$ for some x_2. When $\widehat{g}$ is strict, the last case is forbidden, so that the restriction of $\widehat{g}$ determines a map $|g|_\perp : |G_1|_\perp \to |G_2|_\perp$.

Definition 13. *Let* $\widehat{g} = (g, a) : \widehat{G}_1 \to \widehat{G}_2$ *be a morphism of partially attributed structures. Then* $\widehat{g}$ *(or* g*) is injective if* $|g| : |G_1| \to |G_2|$ *is injective. Assume that* $\widehat{g}$ *is strict, then* $\widehat{g}$ *is surjective on non-attributed elements if* $|g|_\perp : |G_1|_\perp \to |G_2|_\perp$ *is surjective. Besides,* $\widehat{g}$ *preserves attributes if* $\widehat{G}_1$ *and* $\widehat{G}_2$ *are attributed over the same* A *and* $a = id_A$.

As in [7], we assume that all horizontal arrows in the rules preserve attributes, and we will see that this implies that all horizontal arrows in the rewrite steps also preserve attributes. This implies that there are objects A, A_0 and A' and arrows $a : A \to A_0$ and $a' : A_0 \to A'$ in **A** such that, in each rewrite step diagram, the vertical arrows in the top squares are over a and the vertical arrows in the bottom squares are over a'. Let $t_A = a' \circ a : A \to A'$. Typically, elements of A are terms with variables, A' describes types (for example, it could be the final algebra in which carrier sets are singletons) and the morphism t_A gives a type to each variable, and the morphism a denotes an instantiation of the variables (and terms) in A which respects their types:

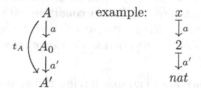

The definitions of PB-PO attributed rewrite rules, matches and steps must ensure that the result of a step is indeed a well-formed attributed structure. Therefore we have to impose some restrictions on rules and matches with respect to re-attribution (i.e. change of attribute value): (i) only items that are explicitly preserved by the rule can be re-attributed; (ii) items being re-attributed can not be identified with anything neither by the match nor by t_L; (iii) the bottom span of the rule must agree with the upper span with respect to re-attribution (for example, it is not possible that the attribute of the bottom span of an item – its type – is changed and the value of the item in the upper span remains unchanged); and (iv) the left- and right-hand sides of the spans of a rule must be fully-attributed. Some of these conditions are defined for the rule and some for the match. Examples 17 and 18 motivate these conditions and illustrate re-attribution issue.

Definition 14 (PB-PO **attributed rewrite rules**). *Given a morphism $t_A :$ $A \to A'$ of **A**, a* PB-PO *attributed rewrite rule over t_A is a* PB-PO *rewrite rule ρ in the category **PAttG** of partially attributed structures, i.e., a diagram:*

$$
\begin{array}{ccccc}
\widehat{L} & \xleftarrow{\;\widehat{l}\;} & \widehat{K} & \xrightarrow{\;\widehat{r}\;} & \widehat{R} \\
\downarrow{\widehat{t}_L} & = & \downarrow{\widehat{t}_K} & = & \downarrow{\widehat{t}_R} \\
\widehat{L}' & \xleftarrow{\;\widehat{l}'\;} & \widehat{K}' & \xrightarrow{\;\widehat{r}'\;} & \widehat{R}'
\end{array}
$$

with the following restrictions: the top line is attributed over A, the bottom line is attributed over A', $\widehat{l}$, $\widehat{l}'$, $\widehat{r}$, and $\widehat{r}'$ are attribute preserving, the vertical morphisms are attributed over t_A, the objects $\widehat{L}$, $\widehat{R}$, $\widehat{L}'$ and $\widehat{R}'$ are totally attributed and the morphism $\widehat{t}_K : \widehat{K} \to \widehat{K}'$ is strict and injective on non-attributed items.

The following condition ensures that whenever an item will be re-attributed, it is (the image of) an item that is preserved by the rule.

Definition 15 (**re-attribution condition**). *Given a* PB-PO *attributed rewrite rule ρ (with notations as above), a totally attributed structure $\widehat{G}$ and a* PB-PO *match $(\widehat{m}, \widehat{m}')$ of ρ in $\widehat{G}$, the match satisfies the* re-attribution condition *with respect to ρ if:*

for each $\underline{n}_G$ in $|G|$, if there is some $\underline{n}_{K'}$ in $|K'|_{\perp}$ with $m'(\underline{n}_G) = l'(\underline{n}_{K'})$ then there is an $\underline{n}_K$ in $|K|$ with $\underline{n}_G = m(l(\underline{n}_K))$ and $\underline{n}_{K'} = t_K(\underline{n}_K)$.

Definition 16 (PB-PO **attributed rewrite system**). *Given a* PB-PO *attributed rewrite rule ρ and a totally attributed structure $\widehat{G}$, a* PB-PO *attributed match of ρ in $\widehat{G}$ is a* PB-PO *match of ρ in $\widehat{G}$ in the category **PAttG**, i.e., a pair $(\widehat{m}, \widehat{m}') = ((m, a), (m', a'))$ such that $\widehat{m} \circ \widehat{m}' = \widehat{t}_L$, with the following restrictions: $\widehat{m}$ is injective and $(\widehat{m}, \widehat{m}')$ satisfies the re-attribution condition with respect to ρ. The* PB-PO *attributed rewrite step applying a* PB-PO *attributed rewrite rule ρ to a* PB-PO *attributed match $(\widehat{m}, \widehat{m}')$ is the* PB-PO *rewrite step applying ρ to $(\widehat{m}, \widehat{m}')$ in the category **PAttG**.*

Example 17. All examples are diagrams having the shape of Diagram (2). We start with two basic examples of re-attribution.

(1) Identity: $\underline{n}$ is preserved, with its attribute.
(2) Identity of structure only: $\underline{n}$ is preserved, but its type and attribute are changed.

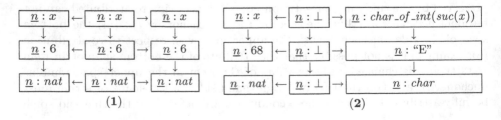

$$(1) \qquad\qquad\qquad\qquad (2)$$

Example 18. We now give three examples to motivate the restrictions about rules and matches made in Definitions 14 and 16.

(1) Here $\hat{t}_K$ is not strict (and thus the rule is not well-formed). The pushout of $(\hat{r}, \hat{n})$ does not exist, so that the rewrite step cannot be constructed.

(2) Here $\hat{t}_K$ is not injective on non-attributed items: again, the rule is not well-formed, and the pushout of $(\hat{r}, \hat{n})$ does not exist.

(3) Here the issue is that $(\hat{m}, \hat{m}')$ does not satisfy the re-attribution condition, since $\underline{n}'$ (in G) should be re-attributed but it is not an item preserved by the rule (not an image of an element in K). The pushout of $(\hat{r}, \hat{n})$ exists, but the resulting H is not totally attributed.

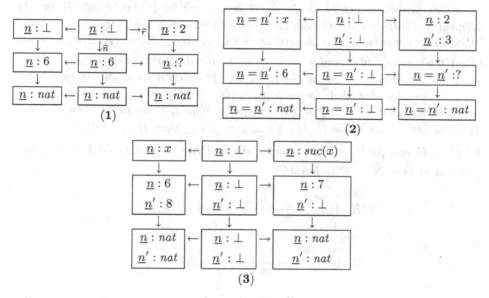

We will use "PB-PO-A" for "PB-PO attributed". In the rest of this section we show that the restrictions imposed on attributed rules and matches in Definitions 14, 15 and 16 are sufficient to guarantee that a rewriting step can be completed, and that the resulting structure is totally attributed. This result is preceded by two technical lemmas concerning pullbacks and pushouts in **PAttG**, respectively.

Lemma 19 (on pullbacks in PAttG). *Let $\hat{G} \xrightarrow{\hat{m}'} \hat{L}' \xleftarrow{\hat{l}'} \hat{K}'$ be a cospan in* **PAttG**, *with $\hat{G}$ and $\hat{L}'$ totally attributed and $\hat{l}'$ attribute-preserving. Let us denote $\hat{G} = (G, A_0, \alpha_G)$, $\hat{L}' = (L', A', \alpha_{L'})$, $\hat{K}' = (K', A', \alpha_{K'})$, $\hat{m}' = (m', a')$ and $\hat{l}' = (l', id_{A'})$, as in the diagram below. Let $G \xleftarrow{g} D \xrightarrow{n'} K'$ be the pullback of $G \xrightarrow{m'} L' \xleftarrow{l'} K'$ in* **G**. *Let $\alpha_D : |D| \rightharpoonup |A_0|$ be the partial map such that, for each $\underline{n}_D \in |D|$: if $|n'|(\underline{n}_D) : \bot$ then $\underline{n}_D : \bot$, otherwise $\underline{n}_D : x_0$ where x_0 is the attribute of $|g|(\underline{n}_D)$. Let $\hat{D} = (D, A_0, \alpha_D)$, $\hat{g} = (g, id_{A_0})$ and $\hat{n}' = (n', a')$.*

Then $\widehat{g} : \widehat{D} \to \widehat{G}$ and $\widehat{n}' : \widehat{D} \to \widehat{K}'$ are morphisms in **PAttG**, $\widehat{n}'$ is strict, and $\widehat{G} \xleftarrow{\widehat{g}} \widehat{D} \xrightarrow{\widehat{n}'} \widehat{K}'$ is the pullback of $\widehat{G} \xrightarrow{\widehat{m}'} \widehat{L}' \xleftarrow{\widehat{l}'} \widehat{K}'$ in **PAttG**.

$$
\begin{array}{ccc}
(G, A_0, \alpha_G) & \xleftarrow{\;(g, id_{A_0})\;} & (D, A_0, \alpha_D) \\
{\scriptstyle (m', a')} \downarrow & PB & \downarrow {\scriptstyle (n', a')} \\
(L', A', \alpha_L) & \xleftarrow{\;(l', id_{A'})\;} & (K', A', \alpha_K)
\end{array}
$$

Lemma 20 (on pushouts in PAttG). *Assume that the functor* $S : \mathbf{G} \to \mathbf{Set}$ *preserves pushouts. Let* $\widehat{D} \xleftarrow{\widehat{n}} \widehat{K} \xrightarrow{\widehat{r}} \widehat{R}$ *be a span in* **PAttG**, *with* $\widehat{R}$ *totally attributed,* $\widehat{r}$ *attribute-preserving and* $\widehat{n}$ *injective, strict, and surjective on non-attributed elements. Let us denote* $\widehat{D} = (D, A_0, \alpha_D)$, $\widehat{K} = (K, A, \alpha_K)$, $\widehat{R} = (R, A, \alpha_R)$, $\widehat{r} = (r, id_A)$ *and* $\widehat{n} = (n, a)$, *as in the diagram below. Let* $D \xrightarrow{h} H \xleftarrow{p} R$ *be the pushout of* $D \xleftarrow{n} K \xrightarrow{r} R$ *in* $\mathbf{G}$. *Then there is a unique total map* $\alpha_H : |H| \to |A_0|$ *such that, for each* $\underline{n}_H \in |H|$: *if* $|p|(\underline{n}_R) = \underline{n}_H$ *for some* $\underline{n}_R : x$ *in* $|R|$ *then* $\underline{n}_H : a(x)$, *and if* $|h|(\underline{n}_D) = \underline{n}_H$ *for some* $\underline{n}_D : x_0$ *in* $|D|$ *then* $\underline{n}_H : x_0$. *Let* $\widehat{H} = (H, A_0, \alpha_H)$, $\widehat{h} = (h, id_{A_0})$ *and* $\widehat{p} = (p, a)$. *Then* $\widehat{H}$ *is totally attributed,* $\widehat{h} : \widehat{D} \to \widehat{H}$ *and* $\widehat{p} : \widehat{R} \to \widehat{H}$ *are morphisms in* **PAttG**, *and* $\widehat{D} \xrightarrow{\widehat{h}} \widehat{H} \xleftarrow{\widehat{p}} \widehat{R}$ *is the pushout of* $\widehat{D} \xleftarrow{\widehat{n}} \widehat{K} \xrightarrow{\widehat{r}} \widehat{R}$ *in* **PAttG**.

$$
\begin{array}{ccc}
(K, A, \alpha_K) & \xrightarrow{\;(r, id_A)\;} & (R, A, \alpha_R) \\
{\scriptstyle (n, a)} \downarrow & PO & \downarrow {\scriptstyle (p, a)} \\
(D, A_0, \alpha_D) & \xrightarrow{\;(h, id_{A_0})\;} & (H, A_0, \alpha_H)
\end{array}
$$

Theorem 21 (rewriting totally attributed structures). *Assume that the functor* $S : \mathbf{AttG} \to \mathbf{Set}$ *preserves pullbacks and pushouts. Then for every* PB-PO-A *rule and every* PB-PO-A *match of this rule, the* PB-PO-A *rewrite step exists, and in addition the resulting* $\widehat{H}$ *is totally attributed.*

Example 22. Let us recall that the rule of Example 2 specifies that the local web pages of the host graph G (i.e., those mapped by m' to node $\underline{l}$ of L') are cloned with all outgoing edges, while edges from the global pages to cloned ones are not copied. Additionally, a selected local node, "root", is linked to a new page.

The following rule intends to enrich the one of Example 2 by specifying that the copy of the local root page should get as attribute the successor of the attribute of the original page ($s : nat \to nat$ is the successor function), and the new page should get in turn its successor.

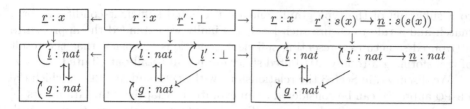

However, if we consider the attributed graph $\widehat{G}$ on the right and the same match as in Example 2 (mapping $\underline{r}_1$, $\underline{l}_1$ and $\underline{l}_2$ to $\underline{l}$ and $\underline{g}_1$ and $\underline{g}_2$ to $\underline{g}$), this match does not satisfy the re-attribution condition, because only $\underline{r}_1$ has a pre-image in K.

$$\widehat{G} = \begin{array}{c} \underline{r}_1:0 \\ \downarrow \\ \underline{l}_1:2 \longrightarrow \underline{l}_2:4 \\ \uparrow \quad \nearrow \quad \downarrow \\ \underline{g}_1:6 \rightleftarrows \underline{g}_2:8 \end{array}$$

The next rule is instead the "right" extension of the rule in Example 2:

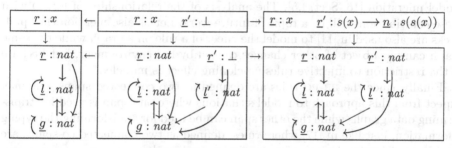

The obvious match satisfies the re-attribution condition, and the resulting $G \xleftarrow{g} D \xrightarrow{h} H$ is:

6   Conclusions and Related Works

We presented a new categorical approach to graph transformation, the PB-PO approach, that combines the standard transformation of structures of the DPO approach with a retyping of the host graph, that allows to model both deletion and cloning of items. PB-PO is shown to be a conservative extension of the AGREE approach, and thus of the SQPO approach with monic matches. The more general framework allows to define a notion of locality parametric with respect to a subgraph of the type graph, and we presented sufficient conditions for a match and a rule to ensure such locality. Finally we extended the approach to attributed structures, presenting sufficient conditions to ensure that the result of transforming a totally attributed structure is still totally attributed.

We discussed in Sect. 3 the relationships with SQPO and AGREE, of which the PB-PO approach can be considered as an evolution. Adaptive Star Grammars [6] were also proposed to model cloning. They include star replacement rules, which can be seen as a restricted kind of DPO rules, and adaptive star rules, which can be applied to arbitrarily large matches via an adaptation mechanism that creates the needed number of copies of items of the lhs. It should be possible to describe this adaptation mechanism with a limit construction, from which one could explore the feasibility of encoding this approach in PB-PO.

Rewriting in the category of spans [15] has been proposed as a framework that generalizes DPO, SPO and SQPO rewriting, thanks to a powerful gluing construction able to model cloning. Transformations based on both pushouts and pullbacks are used in the quite different framework of collagories [12] or that of model migration [16, Sect. 4.5]. The analysis of the relationships of PB-PO with these contributions will be a topic of future work. Let us also mention that pullbacks are also used in [11] to model the effect of a rule on a graph while the same graph can be subject of other changes caused by the environment, but because of the restriction to injective rules no cloning effect is modeled.

Finally, since the PB-PO rules are defined as two connected spans, one may expect from this approach to model situations where one span is used for transforming data graphs while the other span can be used for transforming the typing information, just like in [17] where rules, defined as two connected co-spans, are used to model co-evolutions of meta-models and models.

References

1. Bauderon, M., Jacquet, H.: Node rewriting in graphs and hypergraphs: a categorical framework. Theor. Comput. Sci. **266**(1–2), 463–487 (2001)
2. Cockett, J., Lack, S.: Restriction categories II: partial map classification. Theor. Comput. Sci. **294**(1–2), 61–102 (2003)
3. Corradini, A., Duval, D., Echahed, R., Prost, F., Ribeiro, L.: AGREE – algebraic graph rewriting with controlled embedding. In: Parisi-Presicce, F., Westfechtel, B. (eds.) ICGT 2015. LNCS, vol. 9151, pp. 35–51. Springer, Cham (2015). doi:10.1007/978-3-319-21145-9_3
4. Corradini, A., Heindel, T., Hermann, F., König, B.: Sesqui-pushout rewriting. In: Corradini, A., Ehrig, H., Montanari, U., Ribeiro, L., Rozenberg, G. (eds.) ICGT 2006. LNCS, vol. 4178, pp. 30–45. Springer, Heidelberg (2006). doi:10.1007/11841883_4
5. Corradini, A., Montanari, U., Rossi, F., Ehrig, H., Heckel, R., Löwe, M.: Algebraic approaches to graph transformation - part I: basic concepts and double pushout approach. In: Rozenberg [18], pp. 163–246
6. Drewes, F., Hoffmann, B., Janssens, D., Minas, M.: Adaptive star grammars and their languages. Theor. Comput. Sci. **411**(34–36), 3090–3109 (2010)

7. Duval, D., Echahed, R., Prost, F., Ribeiro, L.: Transformation of attributed structures with cloning. In: Gnesi, S., Rensink, A. (eds.) FASE 2014. LNCS, vol. 8411, pp. 310–324. Springer, Heidelberg (2014). doi:10.1007/978-3-642-54804-8_22
8. Ehrig, H., Heckel, R., Korff, M., Löwe, M., Ribeiro, L., Wagner, A., Corradini, A.: Algebraic approaches to graph transformation - part II: single pushout approach and comparison with double pushout approach. In: Rozenberg [18], pp. 247–312
9. Ehrig, H., Pfender, M., Schneider, H.J.: Graph-grammars: an algebraic approach. In: 14th Annual Symposium on Switching and Automata Theory, Iowa City, Iowa, USA, 15–17 October, pp. 167–180. IEEE Computer Society (1973)
10. Engelfriet, J., Rozenberg, G.: Node replacement graph grammars. In: Rozenberg [18], pp. 1–94
11. Heckel, R., Ehrig, H., Wolter, U., Corradini, A.: Double-pullback transitions and coalgebraic loose semantics for graph transformation systems. Appl. Categorical Struct. 9(1), 83–110 (2001)
12. Kahl, W.: Amalgamating pushout and pullback graph transformation in collagories. In: Ehrig, H., Rensink, A., Rozenberg, G., Schürr, A. (eds.) ICGT 2010. LNCS, vol. 6372, pp. 362–378. Springer, Heidelberg (2010). doi:10.1007/978-3-642-15928-2_24
13. Lack, S., Sobociński, P.: Adhesive categories. In: Walukiewicz, I. (ed.) FoSSaCS 2004. LNCS, vol. 2987, pp. 273–288. Springer, Heidelberg (2004). doi:10.1007/978-3-540-24727-2_20
14. Löwe, M.: Algebraic approach to single-pushout graph transformation. Theor. Comput. Sci. 109(1&2), 181–224 (1993)
15. Löwe, M.: Refined graph rewriting in span-categories - A framework for algebraic graph transformation. In: Ehrig, H., Engels, G., Kreowski, H.-J., Rozenberg, G. (eds.) ICGT 2012. LNCS, vol. 7562, pp. 111–125. Springer, Heidelberg (2012). doi:10.1007/978-3-642-33654-6_8
16. Mantz, F.: Coupled Transformations of Graph Structures applied to Model Migration. Ph.D. thesis, University of Marburg (2014)
17. Mantz, F., Taentzer, G., Lamo, Y., Wolter, U.: Co-evolving meta-models and their instance models: a formal approach based on graph transformation. Sci. Comput. Program. 104, 2–43 (2015)
18. Rozenberg, G. (ed.): Handbook of Graph Grammars and Computing by Graph Transformations, vol. 1: Foundations. World Scientific (1997)
19. Schulz, C., Löwe, M., König, H.: A categorical framework for the transformation of object-oriented systems: models and data. J. Symb. Comput. 46(3), 316–337 (2011)
20. Wouters, L., Gervais, M.P.: Ontology transformations. In: IEEE International Enterprise Distributed Object Computing Conference, pp. 71–80 (2012)

Hierarchical Graph Transformation Revisited
Transformations of Coalgebraic Graphs

Julia Padberg[(⊠)]

Hamburg University of Applied Sciences, Hamburg, Germany
`julia.padberg@haw-hamburg.de`

Abstract. Concepts for structuring are fundamental to any modelling technique. Hierarchical graphs allow vertical structuring, where nodes or edges contain other nodes or subgraphs. There have been several suggestions to hierarchical graphs that differ in terms of the underlying graph type, the elements that are structured and the way the structuring is achieved. In this contribution we aim at a more general notion of hierarchical structures for graphs. We investigate several extensions of the powersets that comprise arbitrarily nested subsets, and call them superpower set. This allows the definition of graphs with possibly infinitely nested nodes. Additionally, we allow edges that are incident to edges. Coalgebras and comma categories are used to capture different notions of hierarchies. The main motivation of this paper is the question how to define recursion on a graph's structure so that we still obtain an $\mathcal{M}$-adhesive transformation system.

Keywords: Graph · Hierarchy · Coalgebra · $\mathcal{M}$-adhesive transformation system

1 Motivation

Graphs are commonly used to describe complex structures that may become very large. For the sake of scalability many approaches using graphs have one or more additional structuring notions. Hierarchical graphs (and graph transformations) add some hierarchy to the nodes or to the edges. Various approaches to graphs with hierarchy have been proposed, e.g. [5,7–9,19,22,24,28,29]. The resulting techniques were used for modelling hierarchical hypermedia, distributed project management, mobile and ubiquitous systems among others.

In this contribution we investigate the possible variations with respect to their use in graph transformation systems. This allows choosing an adequate hierarchical model and directly obtaining the results from $\mathcal{M}$-adhesive transformation systems. The concept of graphs is very general and can be specialised to directed and undirected graphs, as well as hypergraphs, typed, labelled or attributed graphs. To cope with that many different approaches $\mathcal{M}$-adhesive transformation systems offer an abstract definition that requires some categorical constructs and provides a common theory for many different types of graphs and the corresponding transformation systems.

© Springer International Publishing AG 2017
J. de Lara and D. Plump (Eds.): ICGT 2017, LNCS 10373, pp. 20–35, 2017.
DOI: 10.1007/978-3-319-61470-0_2

Hierarchies are naturally presented by means of a recursion, since items of one hierarchical level are expanded to several items of the next lower level. In Sect. 2 we represent the recursive structure of the hierarchy as two constructions that allow nested subsets of subsets. This construction – the superpower set – yields the corresponding functors. Subsequently we investigate these functors and show that coalgebraic graphs based on these functors yield $\mathcal{M}$-adhesive transformation systems. One advantage is that the additional graph structure is not represented by additional arcs but in terms of containment given in a uniform way based on functors. Moreover, we expand the notion of edges so that edges between edges are possible. Although this leads to unconventional concepts such edges have already been used, e.g. in AGG [2,14] or for graph grouping [19]. We investigate several notions of hierarchies in graphs in more detail in Sect. 5:

Hierarchical Graphs comprise many different approaches to hierarchies in graph, in Sect. 5.1 we discuss the relation to three of them, hierarchical hyper-edges as in [9] and packages in [8]. The latter is a more general approach that allows various types of graphs. In [24] hierarchical graphs are given that allow trees as hierarchies for both nodes and edges.

Multi-Hierarchical Graphs as presented in [29] are sketched in Sect. 5.2. Their hierarchy concept is given for nodes and they support several different hierarchies in the same graph that are independent of each other.

Bigraphs [22] are an important theory for modelling ubiquitous computing and combine two graphs to model both locality and connectivity. The first states containment in terms of a forest, the latter the connection via hyperedges. They can be considered to be a special case of hierarchical graphs (see Sect. 5.3).

Graph Grouping as given in [19] allows the analysis of large graphs by group-ing nodes and edges into super nodes and super edges, see Sect. 5.4.

The main benefit of this contribution is the systematic representation of var-ious hierarchy concepts for graphs that support graph transformation in terms of $\mathcal{M}$-adhesive transformation systems, i.e. the algebraic double pushout (DPO) approach, see Sect. 3. Since the hierarchy concepts are the main focus we only have investigated graph types without any additional data as labels or attributes. Section 4 gives an overview over the hierarchy concepts that are obtained as coal-gebraic graphs by varying the underlying categorical concepts. It summarizes the concepts at an informal level and facilitate the establishment of transformation systems for non-standard hierarchical graphs. Then we examine in Sect. 5 var-ious different approaches to hierarchies in graphs. This contribution does not aim at a general theory of hierarchical graphs but at the availability of algebraic graph transformations for widely spread graph hierarchy concepts.

Structure of the paper. In the next section (Sect. 2) we investigate the different constructions for an iterated power set construction. Section 3 details the use of coalgebras for the construction of $\mathcal{M}$-adhesive categories. The emerging hierar-chical concepts are sketched in Sect. 4. We then discuss various approaches to hierarchical graphs (see Sect. 5) and exemplarily compare them to their formu-lation in terms of $\mathcal{M}$-adhesive categories in Sect. 4. Subsequently in Sect. 6, we discuss related work and concluding remarks in Sect. 7 close this contribution.

2   Extension of the Powerset

One of the main technical results of this contribution is given in this section. We introduce superpower sets[1] that allow arbitrary nesting of subsets and are the basic construction for the hierarchy concepts.

The superpower set is achieved by recursively inserting subsets of the super-power set into itself. We here present two possibilities $\mathbb{P}$ and $\mathcal{P}^\omega$. Both are functors that preserve pullbacks of injective morphisms and hence can be deployed in the formulation of $\mathcal{M}$-adhesive transformation systems. These can be constructed from existing ones by various categorical constructions, namely product, coslice, functor and comma categories (see Theorem 4.15 (construction of (weak) adhesive HLR categories) in [10]) and from F-coalgebras based on suitable functors F (see Theorem 1). In Sect. 4 we summarize the potential constructions arising from these results.

The superpower set construction can be defined in various ways, here we present two of them (see the supplemental report [23] for additional ones). Both allow nesting of subsets of arbitrary depth.

Definition 1 (Superpower set $\mathbb{P}$). *Given a set M and $\mathcal{P}(M)$ the power set of M then we define the superpower set $\mathbb{P}(M)$:*

1. *$M \subset \mathbb{P}(M)$ and $\mathcal{P}(M) \subset \mathbb{P}(M)$.*
2. *If $M' \subset \mathbb{P}(M)$ then $M' \in \mathbb{P}(M)$.*

$\mathbb{P}(M)$ is the smallest set satisfying 1. and 2.

Condition 1 ensures that sets may contain nested subset of different depth. $\mathcal{P}^\omega$ differs from $\mathbb{P}$ as in each subset there are only subsets that have the same depth in terms of nesting. So, for some non-empty set M with $m \in M$ we have $\{m, M\} \notin \mathcal{P}^\omega(M)$ but $\{m, M\} \in \mathbb{P}(M)$.

Definition 2 Superpower set $\mathcal{P}^\omega$). *Given a set M we define $\mathcal{P}^0(M) = M$ and $\mathcal{P}^1(M) = \mathcal{P}(M)$ the power set of M. Then $\mathcal{P}^{i+1}(M) = \mathcal{P}(\mathcal{P}^i(M))$ and $\mathcal{P}^\omega(M) = \bigcup_{i \in \mathbb{N}_0} \mathcal{P}^i(M)$.*

The use of strict subsets ensures in both definitions that Russell's antinomy cannot occur. Both superpower set constructions yield well-founded sets with an order based on the depth of the nested parentheses and hence allow induction.

Subsequently, we only investigate the properties of $\mathbb{P}$. The results hold for $\mathcal{P}^\omega$ and other possibilities of superpower sets as well, see [23].

Lemma 1 ($\mathbb{P}$ is a functor). *$\mathbb{P}$: **Sets** $\rightarrow$ **Sets** is defined for sets as in Definition 1 and for functions $f : M \rightarrow N$ by $\mathbb{P}(f) : \mathbb{P}(M) \rightarrow \mathbb{P}(N)$ with*

$$\mathbb{P}(f)(x) = \begin{cases} f(x) & ; x \in M \\ \{\mathbb{P}(f)(x') \mid x' \in x\} & ; \text{ else} \end{cases}$$

---

[1] Although the nesting of sets or nodes (see Sect. 6) are well-known, to the author's knowledge neither the construction nor the corresponding functors have been considered before.

Example 1 (Functor $\mathbb{P}$). Given sets $N_1 = \{u, v, w\}$, $N_2 = \{n_1, n_2, n_3\}$ and $f : N_1 \rightarrow N_2$ with $f : u \mapsto n_1, v \mapsto n_1, w \mapsto n_3$ then we have $\mathbb{P}(f) : \mathbb{P}(N_1) \rightarrow \mathbb{P}(N_2)$ with for example $\mathbb{P}(f)(\{u, v, w, \{u, v\}, \{v, \{w, \emptyset\}\}\}) = \{n_1, n_3, \{n_1\}, \{n_1, \{n_3, \emptyset\}\}\}$.

Lemma 2 ($\mathbb{P}$ preserves injections). *Given injective function $f : M \rightarrow N$ then $\mathbb{P}(f) : \mathbb{P}(M) \rightarrow \mathbb{P}(N)$ is injective.*

For the proof see Lemma 2 in [23]. In this paper the proofs are only given if they are relevant for understanding the concepts.

Lemma 3 ($\mathbb{P}$ preserves pullbacks along injective morphisms). *Given a pullback diagram (PB) in **Sets** with injective $g_1 : C \hookrightarrow D$, then $\mathbb{P}(A)$ in the diagram (1) is pullback in **Sets** as well.*

Proof. Pullbacks and the superpower set functor (see Lemma 2) preserve injections, so $\pi_B : A \hookrightarrow B$, $\mathbb{P}(\pi_B) : \mathbb{P}(A) \hookrightarrow \mathbb{P}(B)$ and $\pi_{\mathbb{P}(B)} : P \hookrightarrow \mathbb{P}(B)$ are injective. Since (PB) is a pullback diagram we have $A = \{(b, c) \mid f_1(b) = g_1(c)\}$. (1) commutes, since $\mathbb{P}$ is a functor. Let P be the pullback of $(\mathbb{P}(D), \mathbb{P}(f_1), \mathbb{P}(g_1))$, so $\mathbb{P}(f_1) \circ \pi_{\mathbb{P}(B)} = \pi_{\mathbb{P}(C)} \circ \mathbb{P}(g_1)$. Hence, $P = \{(B', C') \mid \mathbb{P}(f_1)(B') = \mathbb{P}(g_1)(C')\} \subseteq \mathbb{P}(B) \times \mathbb{P}(C)$.

Moreover, there is the unique $h : \mathbb{P}(A) \rightarrow P$ s.t. $h(A') = (\mathbb{P}(\pi_B)(A'), \mathbb{P}(f\pi_C)(A'))$ for all $A' \subseteq A$ so that the diagrams (2) and (3) commute along h: $\pi_{\mathbb{P}(B)} \circ h = \mathbb{P}(\pi_B)$ and $\pi_{\mathbb{P}(C)} \circ h = \mathbb{P}(\pi_C)$.

We define $\bar{h} : P \rightarrow \mathbb{P}(A)$ with

$$\bar{h}((X, Y)) = \begin{cases} (b, c) \; ; \; \text{if } X = b \in B, \; Y = c \in C \\ \{(x, y) \mid x \in X \cap B, \; y \in Y \cap C, \; f_1(x) = g_1(y)\} \; \cup \\ \qquad \bigcup_{(X', Y') \in (X-B) \times (Y-C)} \bar{h}(X', Y') \; ; \; \text{else} \end{cases}$$

and have:

1. $\bar{h}$ is well-defined since $\bar{h}(X, Y) \in \mathbb{P}(A)$.
2. (2) commutes along $\bar{h}$, i.e. $\mathbb{P}(\pi_B) \circ \bar{h} = \pi_{\mathbb{P}(B)}(X, Y)$ is shown by induction over the depth of the nested parentheses n. For $n = 0$, i.e. the atomic nodes, let $(b, c) \in P$ with $b \in B$ and $c \in C$ be given. $\mathbb{P}(\pi_B) \circ \bar{h}(b, c) = \mathbb{P}(\pi_B)(b, c) = b = \pi_{\mathbb{P}(B)}(b, c)$. Let be $\mathbb{P}(\pi_B) \circ \bar{h}(X, Y) = \pi_{\mathbb{P}(B)}(X, Y)$ for sets with at most n nested parentheses. Given $(\hat{X}, \hat{Y}) \in P$ with $n + 1$ nested parentheses and let $\hat{X} = \hat{B} \cup X$ with $\hat{B} \subseteq B$ and $X \cap B = \emptyset$. Let $\hat{Y} = \hat{C} \cup Y$ with $\hat{C} \subseteq C$ and $Y \cap C = \emptyset$. X and Y have at most n nested parentheses. Then

$$\mathbb{P}(\pi_B) \circ \bar{h}(\hat{X}, \hat{Y})$$

$$= \{x \mid x \in \hat{B} \wedge \exists y \in \hat{C} : f_1(x) = g_1(y)\} \cup \bigcup_{(X',Y') \in (X \times Y)} \mathbb{P}(\pi_B) \circ \bar{h}(X', Y')$$

$$\overset{IB}{=} \hat{B} \cup \bigcup_{(X',Y') \in (X \times Y)} \pi_{\mathbb{P}(B)}(X', Y')$$

$$= \hat{B} \cup X$$

$$= \pi_{\mathbb{P}(B)}(\hat{X}, \hat{Y})$$

$\mathbb{P}(A)$ is isomorphic to the pullback P, since

- $h \circ \bar{h} = id_P$, since $\pi_{\mathbb{P}(B)}$ is injective and $\pi_{\mathbb{P}(B)} \circ h \circ \bar{h} = \mathbb{P}(\pi_B) \circ \bar{h} = \pi_{\mathbb{P}(B)} \circ id_P$.
- $\bar{h} \circ h = id_{\mathbb{P}(A)}$, since $\mathbb{P}(\pi_B)$ is injective and $\mathbb{P}(\pi_B) \circ \bar{h} \circ h = \pi_{\mathbb{P}(B)} \circ h = \mathbb{P}(\pi_B) = \mathbb{P}(\pi_B) \circ id_{\mathbb{P}(A)}$.

For a first impression graphs with arbitrarily nested nodes are defined. Note, only the nodes are nested, but nodes containing others do not have a name. In Fig. 1 the node $\{m_1, m_2, m_3\}$ contains the nodes m_1, m_2 and m_3 but it does not have a name. Edges are hyperedges given as a subset of the superpower set, so the first level of the nesting of subsets defines the edges, so the edge a_4 connects the nodes

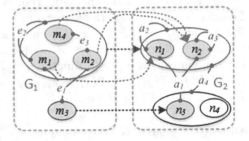

Fig. 1. Morphism in $< Id_{\mathbf{Sets}} \downarrow \mathbb{P} >$

$\{n_1, n_2\}$ and $\{n_2, n_4\}$. The edges have neighbours that are atomic nodes or nodes containing nodes and are given by the neighbour function $ngb : E \to \mathbb{P}(N)$. The category of $\mathbb{P}$-graphs is given by a comma category $< Id_{\mathbf{Sets}} \downarrow \mathbb{P} >$. The morphisms are given by mappings of the nodes and arcs $f = (f_N, f_E) : G_1 \to G_2$ with $f_N : N_1 \to N_2$ and $f_E : E_1 \to E_2$ so that (1) commutes, i.e. $\mathbb{P}(f_N) \circ ngb_1 = ngb_2 \circ f_N$.

$$\begin{array}{ccc} E_1 & \xrightarrow{ngb_1} & \mathbb{P}(N_1) \\ f_E \downarrow & (1) & \downarrow \mathbb{P}(f_N) \\ E_2 & \xrightarrow{ngb_2} & \mathbb{P}(N_2) \end{array}$$

Example 2 (P-Graph). Figure 1 illustrates two $\mathbb{P}$-graphs and the morphism based on the mappings $f_N : N_1 \to N_2$ and $f_E : E_1 \to E_2$ with $f_N(m_4) = n_2$, $f_N(m_i) = n_i$ and $f_E(e_i) = a_i$ for $i = 1, 2, 3$. So, $\mathbb{P}(f_N)(\{m_1, m_2, m_4\}) = \{n_1, n_2\}$.

$$G_1 = (N_1, E_1, ngb_1 : E_1 \to \mathbb{P}(N_1)) \text{ with } (N_2, E_2, ngb_2 : E_2 \to \mathbb{P}(N_2))$$

$$\begin{array}{lll} N_1 = \{m_1, .., m_4\} & \text{and } G_2 = \quad \text{with} & N_2 = \{n_1, .., n_4\} \\ E_1 = \{e_1, e_2, e_3\} & & E_1 = \{a_1, ..., a_4\} \\ ngb_1 : e_1 \mapsto \{m_1, m_2, m_3\} & & ngb_2 : a_1 \mapsto \{n_1, n_2, n_3\} \\ \quad e_2 \mapsto \{m_1, \{m_1, m_2, m_4\} & & \quad a_2 \mapsto \{n_1, \{n_1, n_2\} \\ \quad e_3 \mapsto \{m_2, m_4\} & & \quad a_3 \mapsto \{n_2\} \\ & & \quad a_4 \mapsto \{\{n_1, n_2\}, \{n_2, n_4\}\} \end{array}$$

3 $\mathcal{M}$-Adhesive Categories Using Coalgebras

To express that nodes contain nodes again we need a mapping of nodes to super-power set of nodes $cnt : N \to \mathbb{P}(N)$. This is essentially a coalgebra. Coalgebras are often used for specifying the behaviour of systems and data structures that are potentially infinite, for example classes in object-oriented programming, streams and transition systems.

The second main result shows that coalgebras of functors preserving pullbacks along injective morphisms form an $\mathcal{M}$-adhesive category.

An endofunctor $F : \mathbf{Sets} \to \mathbf{Sets}$ gives rise the category of coalgebras $\mathbf{Sets_F}$ with $M \xrightarrow{\alpha_M} F(M)$ – also denoted by (M, α_M) – being the objects and morphisms $f : (M, \alpha_M) \to (N, \alpha_N)$ – called F-homomorphism – so that (1) commutes in $\mathbf{Sets}$ (see [26]).

$$\begin{array}{ccc} M & \xrightarrow{\alpha_M} & F(M) \\ f \downarrow & (1) & \downarrow F(f) \\ N & \xrightarrow{\alpha_N} & F(N) \end{array}$$

Lemma 4 (Pullbacks along injections in Sets$_F$). *Given a functor F : Sets $\to$ Sets that preserves pullbacks along an injective morphism, then* Sets$_F$ *has pullbacks along an injective F-homomorphism.*

For the proof see Lemma 10 in [23].
$\mathcal{M}$-adhesive transformation systems (e.g. [12,25]) are an abstract framework for graph transformations and allow a uniform description of the different notions and results based on a class $\mathcal{M}$ of specific monomorphisms. These well-known results comprise concepts for transformation, local confluence and parallelism, application conditions, amalgamation and so on.

Definition 3 (Class of monomorphisms $\mathcal{M}$). *Let $\mathcal{M}$ be a class of monomorphisms in* Sets *that is PO-PB-compatible, that is:*

1. *Pushouts along $\mathcal{M}$-morphisms exist and $\mathcal{M}$ is stable under pushouts.*
2. *Pullbacks along $\mathcal{M}$-morphisms exist and $\mathcal{M}$ is stable under pullbacks.*
3. *$\mathcal{M}$ contains all identities and is closed under composition.*

According to Property 4.7 in [26] if $f : M \to N$ is injective in **Sets** then f is an F-monomorphism in **Sets$_F$**. Obviously the class of all injective functions $\mathcal{M}_F = \{(A, \alpha_A) \xhookrightarrow{f} (B, \alpha_B) \mid f$ is injective in **Sets**$\}$ is PO-PB-compatible.

Theorem 1 ($\mathcal{M}$-Adhesive Category). *(Sets$_F$, $\mathcal{M}_F$) is an $\mathcal{M}$-adhesive category if F preserves pullbacks along injective morphisms.*

For the proof see proof of Theorem 1 in [23].
This allows $\mathcal{M}$-adhesive transformation systems for various dynamic systems based on coalgebras of functors the preserve pullbacks along injective morphisms. For a first discussion see [23].

Example 3 (Nested nodes). Nested nodes can be constructed using the coalgebra **Coalg$_\mathbb{P}$** based on the superpower set functor $\mathbb{P}$. Given a set N the function $cnt : N \to \mathbb{P}(N)$ gives the nodes contained in a given node. This function yields an $\mathcal{M}$-adhesive category; the category of coalgebras **Coalg$_\mathbb{P}$** over $\mathbb{P} : \mathbf{Sets} \to \mathbf{Sets}$

with the class $\mathcal{M}$ of injective morphisms.

The nesting of nodes can also be defined allowing the different kinds of nesting using some functor F : **Sets** → **Sets**, so we have the contains function cnt : $N \to \mathsf{F}(N)$. This yields an $\mathcal{M}$-adhesive category where G may be one of the (super-)power functors, e.g. $\mathcal{P}$, $\mathcal{P}^{(1,2)}$, $\mathbb{P}$ or $\mathcal{P}^\omega$ or any other functor preserving pullbacks of injections.

We now investigate the nesting of edges, that is neighbours of edges can again be edges. Edges between edges have been already mentioned for the AGG approach [14] and are used as super edges in graph grouping [19]. To define nested edges we need to extend the neighbour function ngb to edges and nodes. So, we have coalgebraic graphs with directed edges that can be considered as a many sorted coalgebra using the functor F : **Sets** × **Sets** → **Sets** × **Sets** with $F(N, E) = (N, E) \xrightarrow{(!,ngb)} (\mathbf{1}, N \times N)$ where $\mathbf{1}$ is the final object and ! the corresponding final morphism, see e.g. [26]. Corollary 2 in [23] extends the result of Theorem 1 to many sorted coalgebras.

Corollary 1. *Let* $\mathbb{F}$: **Sets** × **Sets** → **Sets** × **Sets** *be given where* $\mathbb{F}$ *preserves pullbacks along injections and let* $\mathcal{M}$ *be the class of pairs of injective morphisms* $< f_N, f_E >$. *Then the category of coalgebraic graphs* ($\mathbf{Coalg}_{\mathbb{F}}, \mathcal{M}$) *is an* $\mathcal{M}$-*adhesive category.*

To give an example we define nested hyperedges.

Example 4 (Nested hyperedges). Given a set of nodes N and a set of edge names E and a function yielding the neighbours $ngb : E \to \mathcal{P}(V \uplus E)$.

Then the category of coalgebras $\mathbf{Coalg}_{F_1}$ over F_1 : **Sets** × **Sets** → **Sets** × **Sets** with $F_1(N, E) = (\mathbf{1}, \mathcal{P}(N \uplus E))$ yields the category of graphs with nested hyperedges. The category of coalgebraic graphs with nested hyperedges $\mathbf{Coalg}_{F_1}$ is an $\mathcal{M}$-adhesive category. Analogously, graphs with nested, undirected edges can be defined by the functor $\mathcal{P}^{(1,2)}$ that yields only subsets containing one ore two elements. And graphs with nested, directed edges can be defined by a functor yielding $(N \uplus E) \times (N \uplus E)$, for both see [23].

Subsequently some properties are sketched that might be worthwhile when investigating coalgebraic graphs more deeply. These properties can be defined independently of the underlying constructs and functors. Their definition depends on the purpose of the approach and we merely hint at a possible formulation as this contribution does not aim at a general theory of hierarchical graph transformation.

Properties of nested nodes can be defined for example as follows:

– Nodes names are unique if cnt is injective.
– Nodes referring to themselves are the atomic nodes, so we define the set $aN = \{n \mid cnt(n) = n\}$.
– Nodes are containers if they are not atomic.
– The set of nodes is well-founded if and only if
  – $X \in N \wedge Y \in cnt(X)$ implies, that $Y \in cnt(N)$

– $X \in cnt(N) \wedge Y \in (X - N)$ implies, that $Y \in cnt(N)$
This ensures that the contains function cnt yields a directed acyclic graph.
– The set of nodes is hierarchical if and only if $cnt(n) \cap cnt(n') \neq \emptyset$ implies $n = n'$. This ensures that the neighbour function cnt yields a forest.

Properties of nested edges can be defined for example as follows:

– Common edges are those without nested edges. Hence, the set of common edges is given by those edges e, where $ngb(e)$ does not contain any edge.
– The function $ngb^* : E \to \mathcal{P}(N)$ yields the set of all incident nodes of arbitrarily deep nesting and is defined by $ngb^*(e) = \{n \in N \mid n \in ngb(e)\} \cup \bigcup_{x \in ngb(e)} ngb^*(x)$. Edges are node-based if is ngb^* well-defined. They are not node-based if they are incident to some container node which is not well-founded.
– Edges are atomic if they are node-based and if the function $ngb^*(E) \subseteq aN$ only yields atomic nodes.

The notion of subgraphs given by a set of nodes can be transferred easily to the above concepts using the recursive extension $cnt^*(n) = \{n \in N \mid n \in cnt(n)\} \cup \bigcup_{x \in cnt(n)} cnt^*(x)$. A subgraph $G[M] \subseteq G = (N, E, cnt, ngb)$ induced by a subset of nodes $M \subseteq N$ can be defined by $G[M] = (M, \{e \in E \mid ngb(e) \in cnt^*(M)\}, cnt, ngb)$ where cnt, ngb are the corresponding restrictions. This yields a notion of subgraph that comprises edges between inducing nodes. Note that in [24] a different approach is chosen (see Sect. 5.1) where edges between nodes that have the same parent need not have this parent.

4 $\mathcal{M}$-Adhesive Categories of Hierarchical Graphs

The results of the both previous sections yield different $\mathcal{M}$-adhesive categories depending on the choice of the categorical construction and the underlying functors. We obtain different types of hierarchical graphs and corresponding $\mathcal{M}$-adhesive transformation systems. Hence (DPO) transformations for each of these types of hierarchical graphs are provided.

The neighbours of edges are given by the neighbour function ngb and the nodes contained by node are given by the contains function cnt. The nodes may be containers or atomic nodes, containers may have a name or not, depending on the construction. The edges can be the well-known ones, (un-)directed, hyperedges with or without an order as well as ones having edges between edges. Next, we state the categorical constructions, involved functors and relate the resulting categories to the examples in this paper. For details and proofs see [23].

Example 5 (Types of hierarchical graphs and corresponding $\mathcal{M}$-adhesive transformation systems).

1. The comma category $< Id_{\mathbf{Sets}} \downarrow \mathbb{P} >$ as used in Example 2 is an $\mathcal{M}$-adhesive category because of the comma-category construction (see Theorem 4.15 in [10]) and $\mathbb{P}$ preserving pullbacks of injections. It yields hierarchical graphs with hyperedges between nodes and containers of nodes, but containers do not have an explicit name.

2. Combining the nested nodes based on the superpower set functor $\mathbb{P}$ as in Example 3 with usual edges concepts leads to various types of coalgebraic graphs and is closely related to hierarchical graphs in the sense of [8]. In this case hierarchical graphs are given by $G = (N, E, cnt : N \to \mathbb{P}(N), ngb : E \to \mathsf{H}(N))$. H determines edge type. Typical choices for H are $\mathcal{P}$ or $(\_)^*$ for hyperedges, $\mathcal{P}^{(1,2)}$ for undirected edges or for directed edges the copying functor $\mathsf{X}^2 : \mathbf{Sets} \to \mathbf{Sets} \times \mathbf{Sets}$ with $\mathsf{X}^2(N) = N \times N$. For an example see Sect. 5.1. Item 1.

We use a coalgebra over $\mathbb{F}_1 : \mathbf{Sets} \times \mathbf{Sets} \to \mathbf{Sets} \times \mathbf{Sets}$ with $\mathbb{F}(N, E) = \mathbb{P}(N) \times \mathsf{H}(N)$. Then the category of coalgebraic graphs $(\mathbf{Coalg}_{\mathbb{F}_1}, \mathcal{M})$ is an $\mathcal{M}$-adhesive category.

3. A hierarchy where the edges are refined by subnets (see [9]) is obtained by the neighbouring function $ngb : E \to (N)^* \times \mathcal{P}^\omega(N)$ that maps edges to a pair where the first component defines the incident nodes and the second component defines the nodes contained by the edges. This nesting is layered as it is defined by the functor $\mathcal{P}^\omega$, see Definition 2. The resulting graphs are given by $G = (N, E, ngb : E \to N^* \times \mathcal{P}^\omega(N))$. The category of such graphs is given by the comma category $< Id_{\mathbf{Sets}} \downarrow \mathsf{G} >$ with the functor $\mathsf{G} = ((\_)^* \times \mathcal{P}^\omega) \circ \mathsf{X}^2$.

Note $\mathsf{G}(N) = ((\_)^* \times \mathcal{P}^\omega) \circ \mathsf{X}^2(N) = ((\_)^* \times \mathcal{P}^\omega)(N, N) = N^* \times \mathcal{P}^\omega(N)$. For an example see Sect. 5.1. Item 2.

4. For hierarchies where the edges between nodes may have other parents than the nodes and where the edges may contain subgraphs (as in [24]) the coalgebraic graphs are given by the functions $cnt : N \to \mathbb{P}(N \uplus E)$ and $ngb : E \to \mathcal{P}(N) \times \mathbb{P}(N \uplus E)$. We use the coalgebra with $\mathbb{F}_2(N, E) = (\mathbb{P}(N \uplus E), \mathcal{P}(N) \times \mathbb{P}(N \uplus E))$. $\mathbb{P}(N \uplus E)$ yields nested sets of nodes and edges and $\mathcal{P}(N)$ yields the incident nodes of an hyperedge. To obtain an $\mathcal{M}$-adhesive category $\mathbf{Coalg}_{\mathbb{F}_2}$ we construct $\mathbb{F}_2$ from other functors that yield $\mathcal{M}$-adhesive categories. An example is in Sect. 5.1. Item 3

5. Multiple hierarchies can be constructed as coalgebraic graphs using a copying functor $\mathsf{X}^i : \mathbf{Sets} \to \mathbf{Sets} \times \mathbf{Sets} \times \ldots \times \mathbf{Sets}$. Then the contains function $cnt : N \to \prod \circ \mathsf{X}^i \circ \mathbb{P}(N)$ yields for each node i different nestings. For edges we may use hyperedges $ngb : E \to \mathcal{P}(N)$. The corresponding $\mathcal{M}$-adhesive category $\mathbf{Coalg}_{\mathbb{F}_3}$ of coalgebraic graphs is given by $\mathbb{F}_3(N, E) = (\prod \circ \mathsf{X}^i \circ \mathbb{P}(N), \mathcal{P}(N))$ and corresponds to the multi-hierarchical graphs in Sect. 5.2.

6. For bigraphs, see Sect. 5.3, we use the functions $cnt : N \to \mathbb{P}(N)$ and $ngb : E \to \mathcal{P}(N \uplus E) \times \mathcal{P}(N \uplus E)$. Again we obtain an $\mathcal{M}$-adhesive category $\mathbf{Coalg}_{\mathbb{F}_4}$ of coalgebraic graphs with $\mathbb{F}_4(N, E) = (\mathbb{P}(N), \mathcal{P}(N \uplus E) \times \mathcal{P}(N \uplus E))$ constructed from other functors.

7. The functions $cnt : N \to \mathbb{P}(N)$ and $ngb : E \to N \times N \times \mathbb{P}(E)$ allow the description of graph grouping and give rise to the category of coalgebraic graphs $\mathbf{Coalg}_{\mathbb{F}_5}$ with $\mathbb{F}_5(N, E) = (\mathbb{P}(N), N \times N \times \mathbb{P}(E))$ that corresponds roughly to the the graph grouping in Sect. 5.4.

Varying the constructions, mainly comma categories, product categories and coalgebras and varying the involved functors yield a huge amount of different hierarchy concepts that all lead to well-defined transformation systems. The above examples have been selected to show the width of this approach and to relate it to existing notions of hierarchical graphs. It may as well be used to define new appropriate hierarchy concepts. So, for a specific application the employed hierarchy concept can be chosen out of many different ones.

5 Transformations of Hierarchical Graphs

Here we argue to what extent known concepts can be considered as $\mathcal{M}$-adhesive categories of hierarchical graphs. The detailed, mathematical investigation of each of these examples is beyond the scope of this paper.

Labels and attributes are not considered in this paper, but labelled or attributed graphs yield $\mathcal{M}$-adhesive categories (see [10,12]) and at least labels can be introduced into coalgebraic constructions (see [1,26]).

5.1 Hierarchical Graphs

Many possibilities to define hierarchical graphs have already been investigated, e.g. [4,6–9,24]. In [13] the possibility of infinitely recursive hierarchies has already been introduced as an infinite number of type layers. Here we sketch how three of them, namely [8,9,24], can be considered in this framework.

1. *Hierarchical Graphs as in* [8]

   In this approach graphs are grouped into packages via a coupling graph. A hierarchical graph is a system $H = (G, D, B)$, where G is a graph some graph type, P is a rooted directed acyclic graph, and B is a bipartite coupling graph whose partition contains the nodes of N_G and of N_P. All edges are oriented from the first N_G to the second set of nodes N_P and every node in N_G is connected to at least one node in N_P. For this approach we can consider coalgebraic graphs in the coalgebra category $\mathbf{Coalg}_{\mathbb{F}_1}$ (see Example 5.2) with $cnt : N \to \mathbb{P}(N)$ being well-founded. Additionally a completeness condition, stating that each atomic node is within some package, has to hold:
   $\forall n \in N : cnt(n) = n \Rightarrow \exists p \in N : n \in cnt(p)$

The packages are the nodes that are not atomic. The edge function is given by $ngb : E \to \mathsf{H}(N)$ where $\mathsf{H}(N)$ determines the type of the underlying graphs. In Fig. 2 we have an example with two packages, that uses directed egdes. So based on $\mathsf{H} = \mathsf{X}^2$ we can give this example as a coalgebraic graph.

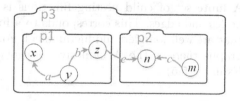

Fig. 2. Hierarchical graph as in [8]

We have $N = \{n, m, x, y, z, p1, p2, p3\}$

with $cnt(v) = \begin{cases} v & \text{; if } v \in \{n, m, x, y, z\} \\ \{x, y, z\} & \text{; if } v = p1 \\ \{n, m\} & \text{; if } v = p2 \\ \{p1, p2\} & \text{; if } v = p3 \end{cases}$ and $ngb : \begin{cases} a \mapsto (y, x) \\ b \mapsto (y, z) \\ c \mapsto (m, n) \\ e \mapsto (z, n) \end{cases}$

2. *Hierarchical Hypergraphs as in* [9] Hypergraphs $H = (V, E, att, lab)$ in [9] consist of two finite sets V and E of vertices and hyperedges. These are equipped with an order, so the attachment function is defined by $att : E \to V^*$. The hierarchy is given in layers, in the sense that subsets in the same layer have the same nesting depth. So, edges are within one layer. Hierarchical graphs $< G, F, cts : F \to \mathcal{H} >\in \mathcal{H}$ are given with special edges F that contain potentially hierarchical subgraphs. Figure 3a depicts a hierarchical graph that can be considered to be a graph in the comma category $< Id_{\mathbf{Sets}} \downarrow \mathsf{G} >$ (see Example 5.3). The graph $G = (N, E, ngb)$ with $ngb : E \to N^* \times \mathcal{P}^\omega(N))$ is defined so that edges are node-based.

3. *Hierarchical Graphs as in* [24] are obtained from hypergraphs by adding a parent assigning function to them. Nodes and edges can be assigned as a child of any other node or edge. These correspond to coalgebraic graphs in the category $\mathbf{Coalg}_{\mathbb{F}_2}$ (see Example 5.4). The parent function coincides with $cnt : N \to \mathbb{P}(N \uplus E)$ being well-founded and hierarchical and $ngb : E \to \mathcal{P}(N) \times \mathbb{P}(N \uplus E)$ since edges can have children as well.

In Fig. 4 the nodes $N = \{1, 2, 3, 6, 8, 9, 11\}$ and the contains function $cnt : N \to \mathbb{P}(N \uplus E)$, yield the nodes and their children. The hyperedges $E = \{4, 5, 7, 10\}$ with $ngb : E \to \mathcal{P}(N \times \mathbb{P}(N \uplus E))$ yield the edges. Note in this example the edges are not nested.
Contains and neighbour function are given by

$cnt : 1 \mapsto 1 \qquad 8 \mapsto 8$
$ 2 \mapsto 2 \qquad 9 \mapsto 9$
$ 3 \mapsto \{1, 2, 4\} \quad 11 \mapsto \{8, 9\}$
$ 6 \mapsto 6$

and $ngb : 4 \mapsto (\{1, 2\}, \emptyset)$
$ 5 \mapsto (\{2, 6\}, \emptyset)$
$ 7 \mapsto (\{3, 6, 11\}, \emptyset)$
$ 10 \mapsto (\{8, 9\}, \emptyset)$

5.2   Multi-Hierarchical Graphs

In [29] multiple hierarchies have been suggested, first ideas can be found in [24]. A finite set of child nesting functions is specified that relate nodes to set of nodes and edges. This corresponds to a finite family $(cnt_i : N \to \mathbb{P}(N \uplus E))_{i<n}$ that are well-founded and hierarchical. For transformations of multi-hierarchical graphs there is the $\mathcal{M}$-adhesive category $\mathbf{Coalg}_{\mathbb{F}_3}$ of coalgebraic graphs (see Example 5.5).

5.3   Bigraphs as an Hierarchy

Bigraphs [22] originate in process calculi for concurrent systems and provide a graphical model of computation. A bigraph is composed of two graphs: a place

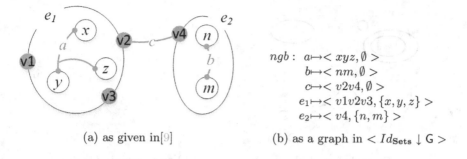

$$ngb:\ a \mapsto\ <xyz, \emptyset>$$
$$b \mapsto\ <nm, \emptyset>$$
$$c \mapsto\ <v2v4, \emptyset>$$
$$e_1 \mapsto\ <v1v2v3, \{x, y, z\}>$$
$$e_2 \mapsto\ <v4, \{n, m\}>$$

(a) as given in[9]   (b) as a graph in $<Id_{\mathsf{Sets}} \downarrow \mathsf{G}>$

Fig. 3. Example of hierarchical hypergraphs

graph and a link graph. They emphasize interplay between physical locality and virtual connectivity. Reaction rules allow the reconfiguration of bigraphs. A bigraphical reactive system consists of a set of bigraphs and a set of reaction rules, which can be used to reconfigure the set of bigraphs. Bigraphs may be composed and have a bisimulation that is a congruence wrt. composition. Categorically, bigraphs are given as morphisms in a symmetric partial monoidal category where the objects are interfaces. This construction corresponds to ranked graphs as given in [15] where morphisms are given by a isomorphism class of concrete directed graphs with interfaces. [11] discusses extensively the relation of bigraphs to graph transformations. In [16] a functor that flattens bigraphs into ranked graphs is provided that encodes the topological structure of the place graph into the node names. In [5] bigraphs are shown to be essentially the same as gs-graphs that present the place and the link graph within one graph. We also represent bigraphs within one graph, where the hierarchical structure is given by a superpower set of nodes and the link structure is given by nested hyperedges. Here we abstract from the categorical foundations and give bigraphs as a special cases of hierarchical graphs. Hence, we ignore their categorical structure,

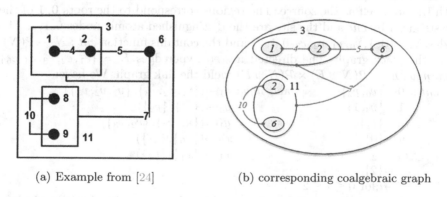

(a) Example from [24]   (b) corresponding coalgebraic graph

Fig. 4. Hierarchical graph in [24]

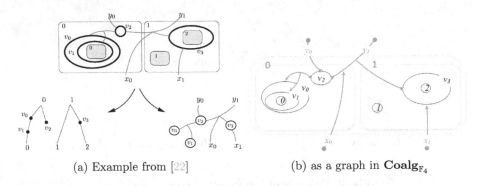

(a) Example from [22]   (b) as a graph in **Coalg**$_{\mathbb{F}_4}$

Fig. 5. Bigraph

but we obtain a transformation system. Nevertheless, often only the graphical representation of bigraphs is used [3,30,31].

A bigraph is a 5-tuple: $(V, E, ctrl, prnt, link) : \langle k, X \rangle \to \langle m, Y \rangle$, where V is a set of nodes, E is a set of edges, $ctrl$ is the control map that assigns controls to nodes, $prnt$ is the parent map that defines the nesting of nodes, and $link$ is the link map that defines the link structure. The notation $\langle k, X \rangle \to \langle m, Y \rangle$ indicates that the bigraph has k holes (sites) and a set of inner names X and m regions, with a set of outer names Y. These are respectively known as the inner and outer interfaces of the bigraph.

Below we illustrate the relation of bigraphs to coalgebraic graphs in **Coalg**$_{\mathbb{F}_4}$ (see Example 5.6) in an example. In Fig. 5a we have an introductory example from [22] that we represent as a coalgebraic graph. The coalgebraic graphs in the $\mathcal{M}$-adhesive category **Coalg**$_{\mathbb{F}_4}$ need to have well-founded and hierarchical nodes, where the contains function represents the parent function, so $cnt = prnt$. The function $cntrl$ yields basically the in- and out-degree of each node. The $link$ function yields hyperedges, which we represent as directed hyperedges. Hyperedges connecting outer names are represented as directed hyperedges with the arc itself as the target, those connecting inner names as directed hyperedges with the arc itself as the source. The regions correspond to the roots $\mathbf{0}, \mathbf{1}$ of the forests given by cnt and the site are the distinguished atomic nodes $0, 1, 2$. The nodes $N = \{\mathbf{0}, \mathbf{1}, v_0, v_1, v_2, v_3, 0, 1, 2\}$ and the contains function $cnt : N \to \mathbb{P}(N)$, yield the place graph. The directed nested hyperedges $E = \{e_1, e_2, e_3, e_4, e_5\}$ with $ngb : E \to \mathbb{P}(N \uplus E) \times \mathbb{P}(N \uplus E)$ yield the link graph. We have:

$$cnt : \ \mathbf{0} \mapsto \{v_0, v_2\} \qquad \text{and } ngb : e_1 \mapsto (\{v_1, v_2, v_3\}, \{v_1, v_2, v_3\})$$
$$\mathbf{1} \mapsto \{v_3, 1\} \qquad\qquad\qquad\quad y_0 \mapsto (\{v_2\}, \{v_2\})$$
$$v_0 \mapsto \{v_1\} \qquad\qquad\qquad\quad y_1 \mapsto (\{v_2, v_3\}, \{v_2, v_3\})$$
$$v_1 \mapsto \{0\} \qquad\qquad\qquad\quad x_0 \mapsto (\{x_0\}, \{y_1\})$$
$$v_2 \mapsto v_2 \qquad\qquad\qquad\quad x_1 \mapsto (\{x_1\}, \{v_3\})$$
$$v_3 \mapsto \{2\}$$
$$i \mapsto i; \text{for } 0 \leq i \leq 2$$

Assuming cnt to be just well-founded we obtain bigraphs with sharing as in [28].

5.4   Graph Grouping

[19] aims at a fundamentally different application area, namely graph grouping to support data analysts making decisions based on very large graphs. Here, a graph hierarchy is established to cope with large amounts of data and to aggregate them. Graph grouping operators produce a so-called summary graph containing super vertices and super edges. A super vertex stores the properties representing the group of nodes, and a super edge stores the properties representing the group of edges. Basically this leads to a contains function $cnt : N \rightarrow \mathbb{P}(N)$ that are well-founded but not necessarily hierarchical and a neighbour function $ngb : E \rightarrow N \times N \times \mathbb{P}(E)$. These can be given as coalgebraic graphs in the category of coalgebras $\mathbf{Coalg}_{\mathbb{F}_5}$ (see Example 5.7) that is $\mathcal{M}$-adhesive.

But clearly this graph grouping is only sensible for attributed graphs since these used to abstract the data.

6   Related Work

Abstraction in graph transformations is employed for different purposes, e.g. for model checking, for a common theory for different types of graphs, for transferring concepts and results. Abstract approaches to graph transformations [10,12] of different types of graphs comprise mainly $\mathcal{M}$-adhesive transformation systems. Other approaches to abstract graphs can be found in that uses the presentation of graphs as a comma-category [17,27] or as a coalgebra [18]. F-graphs are a family of graph categories induced by a comma category construction using a functor F (see [27]). In [17] the notion of F-graphs based on a construction that is a comma-category and has been encoded as a coalgebraic construction in [18] using the basic idea from [26].

In [20,21] coalgebraic signatures are used to define various graph types and yields first steps towards a new paradigm for graph transformation systems. Moreover [20] is concerned with attributed graph transformations, since the coalgebraic definition allows a uniform treatment of term algebras over arbitrary signatures and unstructured label sets. in contrast this contribution is concerned with the inner structure of hierarchical graphs and establishes coalgebras as another possible construction for $\mathcal{M}$-adhesive transformation systems.

7   Concluding Remarks

We have presented a novel approach to hierarchies in graphs and graph transformations. This approach supports the use of the mature and extensive theory of algebraic graph transformations for graphs with many different and also uncommon hierarchy concepts. The aim of our approach is not a generalisation of hierarchy concepts in graph transformation but a possibility to access algebraic graph transformation for graphs with a wide spectrum of hierarchy concepts.

The vision is a clear and simple access that provides a potential user with the hierarchical technique that is most adequate for the purpose. This requires a much deeper treatment of the hierarchical concepts at the abstract categorical level as well as an intuitive representation of these concepts.

To aim at this vision future work comprises then formulation of typical results and notions for hierarchical graph transformations, as e.g. flattening, hierarchical rule application or imposing a hierarchy. Moreover, the inclusion of labels, types and attributes is central for realizing that vision. For this task the work in [21] is an exciting prospect. Additionally, the transfer of existing concepts to this more categorical approach is required based on further investigation of the relations discussed in Sect. 5.

Acknowledgements. I am very grateful for the constructive and thorough comments of the anonymous referees.

References

1. Adamek, J.: Introduction to coalgebra. Theor. Appl. Categories **14**, 157–199 (2005). http://www.tac.mta.ca/tac/volumes/14/8/14-08abs.html
2. AGG: The attributed graph grammar system (2014). http://user.cs.tu-berlin.de/~gragra/agg/, revision: 10/29/2014 16:43:00
3. Benford, S., Calder, M., Rodden, T., Sevegnani, M.: On lions, impala, and bigraphs: Modelling interactions in physical/virtual spaces. ACM Trans. Comput. Hum. Interact. **23**(2), 9: 1–9: 56 (2016). http://doi.acm.org/10.1145/2882784(2016)
4. Bruni, R., Corradini, A., Montanari, U.: Modeling a service and session calculus with hierarchical graph transformation. ECEASST 30 (2010). http://journal.ub.tu-berlin.de/index.php/eceasst/article/view/427
5. Bruni, R., Montanari, U., Plotkin, G.D., Terreni, D.: On hierarchical graphs: reconciling bigraphs, Gs-monoidal theories and Gs-graphs. Fundam. Inform. **134**(3–4), 287–317 (2014). http://dx.doi.org/10.3233/FI-2014-1103
6. Busatto, G.: An abstract model of hierarchical graphs and hierarchical graph transformation. Ph.D. thesis, University of Paderborn, Germany (2002). http://ubdata.uni-paderborn.de/ediss/17/2002/busatto/disserta.pdf
7. Busatto, G., Hoffmann, B.: Comparing notions of hierarchical graph transformation. Electr. Notes Theor. Comput. Sci. **50**(3), 310–317 (2001). http://dx.doi.org/10.1016/S1571-0661(04)00184--7
8. Busatto, G., Kreowski, H., Kuske, S.: Abstract hierarchical graph transformation. Math. Struct. Comput. Sci. **15**(4), 773–819 (2005). http://dx.doi.org/10.1017/S0960129505004846
9. Drewes, F., Hoffmann, B., Plump, D.: Hierarchical graph transformation. J. Comput. Syst. Sci. **64**(2), 249–283 (2002). http://dx.doi.org/10.1006/jcss.2001.1790
10. Ehrig, H., Ehrig, K., Prange, U., Taentzer, G.: Fundamentals of Algebraic Graph Transformation. EATCS Monographs in TCS. Springer (2006)
11. Ehrig, H.: Bigraphs meet double pushouts. Bull. ATCS **78**, 72–85 (2002)
12. Ehrig, H., Golas, U., Hermann, F.: Categorical frameworks for graph transformation and HLR systems based on the DPO approach. Bull. EATCS **102**, 111–121 (2010)
13. Engels, G., Schürr, A.: Encapsulated hierarchical graphs, graph types, and meta types. Electr. Notes Theor. Comput. Sci. **2**, 101–109 (1995). http://dx.doi.org/10.1016/S1571-0661(05)80186--0

14. Ermel, C., Rudolf, M., Taentzer, G.: The agg approach: language and environment. In: Ehrig, H., Engels, G., Kreowski, H.J., Rozenberg, G. (eds.) Handbook of Graph Grammars and Computing by Graph Transformation, pp. 551–603. World Scientific Publishing Co., Inc. (1999). http://dl.acm.org/citation.cfm?id=328523. 328619

15. Gadducci, F., Heckel, R.: An inductive view of graph transformation. In: Presicce, F.P. (ed.) WADT 1997. LNCS, vol. 1376, pp. 223–237. Springer, Heidelberg (1998). doi:10.1007/3-540-64299-4_36

16. Gassara, A., Rodriguez, I.B., Jmaiel, M., Drira, K.: Encoding bigraphical reactive systems into graph transformation systems. Electron. Notes Discrete Math. **55**, 207–210 (2016). http://dx.doi.org/10.1016/j.endm.2016.10.051

17. Jäkel, C.: A unified categorical approach to graphs (2015). https://arxiv.org/abs/1507.06328

18. Jäkel, C.: A coalgebraic model of graphs (2016). https://arxiv.org/abs/1508.02169

19. Junghanns, M., Petermann, A., Rahm, E.: Distributed grouping of property graphs with gradoop. In: Proceedings of the 17. Fachtagung, Datenbanksysteme für Business, Technologie und Web. LNI, GI (2017) (to be published)

20. Kahl, W.: Categories of coalgebras with monadic homomorphisms. In: Bonsangue, M.M. (ed.) CMCS 2014 2014. LNCS, vol. 8446, pp. 151–167. Springer, Heidelberg (2014). doi:10.1007/978-3-662-44124-4_9

21. Kahl, W.: Graph transformation with symbolic attributes via monadic coalgebra homomorphisms. ECEASST **71** (2014). http://journal.ub.tu-berlin.de/eceasst/article/view/999

22. Milner, R.: Pure bigraphs: structure and dynamics. Inf. Comput. **204**(1), 60–122 (2006). http://dx.doi.org/10.1016/j.ic.2005.07.003

23. Padberg, J.: Towards $\mathcal{M}$-adhesive categories of coalgebraic graphs. Technical report, ArXiv e-prints (2017). https://arxiv.org/abs/1702.04650

24. Palacz, W.: Algebraic hierarchical graph transformation. J. Comput. Syst. Sci. **68**(3), 497–520 (2004). http://dx.doi.org/10.1016/S0022-0000(03)00064--3

25. Prange, U., Ehrig, H., Lambers, L.: Construction and properties of adhesive and weak adhesive high-level replacement categories. Appl. Categorical Struct. **16**(3), 365–388 (2008)

26. Rutten, J.: Universal coalgebra: a theory of systems. Theor. Comput. Sci. **249**(1), 3–80 (2000). http://www.sciencedirect.com/science/article/pii/S0304397500000566

27. Schneider, H.J.: Describing systems of processes by means of high-level replacement. In: Handbook of Graph Grammars and Computing by Graph Transformation, vol. 3, pp. 401–450. World Scientific (1999)

28. Sevegnani, M., Calder, M.: Bigraphs with sharing. Theor. Comput. Sci. **577**, 43–73 (2015). http://dx.doi.org/10.1016/j.tcs.2015.02.011

29. Ślusarczyk, G., Łachwa, A., Palacz, W., Strug, B., Paszyńska, A., Grabska, E.: An extended hierarchical graph-based building model for design and engineering problems. Autom. Const. **74**, 95–102 (2017)

30. Walton, L.A., Worboys, M.: A qualitative bigraph model for indoor space. In: Xiao, N., Kwan, M.-P., Goodchild, M.F., Shekhar, S. (eds.) GIScience 2012. LNCS, vol. 7478, pp. 226–240. Springer, Heidelberg (2012). doi:10.1007/978-3-642-33024-7_17

31. Worboys, M.F.: Using bigraphs to model topological graphs embedded in orientable surfaces. Theor. Comput. Sci. **484**, 56–69 (2013). http://dx.doi.org/10.1016/j.tcs.2013.02.018

Geometric Modeling: Consistency Preservation Using Two-Layered Variable Substitutions

Thomas Bellet[1](✉), Agnès Arnould[2], Hakim Belhaouari[2], and Pascale Le Gall[1]

[1] MICS, CentraleSupélec, University of Paris-Saclay, Paris, France
{thomas.bellet,pascale.legall}@centralesupelec.fr
[2] XLIM UMR CNRS 7252, University of Poitiers, Poitiers, France
{agnes.arnould,hakim.ferrier.belhaouari}@univ-poitiers.fr

Abstract. In the context of topology-based geometric modeling, operations transform objects regarding both their topological structure (*i.e.* cell subdivision: vertex, edge, face, etc.) and their embeddings (*i.e.* relevant data: vertex positions, face colors, volume densities, etc.). Graph transformations with variables allow us to generically handle those operations. We use two types of variables: orbit variables to abstract topological cells and node variables to abstract embedding data. We show how these variables can be simultaneously used, and we provide syntactic conditions on rules to ensure that they preserve object consistency. This rule-based approach is the cornerstone of Jerboa, a tool that allows a fast and safe prototyping of geometric modelers.

Keywords: DPO graph transformations · Labeled graphs · Graph variables · Topology-based geometric modeling · Consistency preservation

1 Introduction

Context. Geometric modeling concerns mathematical models useful to create, manipulate, modify or display realistic n-dimensional (nD) objects in numerous application domains such as computer-aided design and manufacturing, mechanical engineering, architecture, geology, archaeology, medical image processing, scientific visualization, animated movies or video games. Many modeling tools are therefore developed at expensive costs to fulfill the various application needs. In order to facilitate the prototyping of new modelers, we developed a tool set for designing and generating safe geometric modeler kernels, called Jerboa[1] [1].

The Jerboa Tool Set. Objects are specified as generalized maps (G-maps) [7] which allow uniform modeling of complex nD objects (*e.g.* 2D surfaces, 3D volumes) by regular graphs. Their topological structure (*i.e.* cell subdivision) is encoded by the graph structure and the arc labels, while their embeddings (*i.e.* geometric or applicative data) are given by the node labels. Their consistency is guaranteed thanks to G-map labeling constraints. Designing a modeler therefore

---

[1] http://xlim-sic.labo.univ-poitiers.fr/jerboa/.

© Springer International Publishing AG 2017
J. de Lara and D. Plump (Eds.): ICGT 2017, LNCS 10373, pp. 36–53, 2017.
DOI: 10.1007/978-3-319-61470-0_3

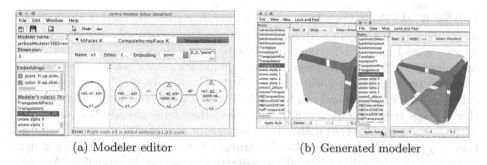

(a) Modeler editor   (b) Generated modeler

Fig. 1. Design and application of the face triangulation with Jerboa

starts by specifying the dimension and the embeddings of manipulated objects (*e.g.* 3D objects with vertex positions and face colors in Fig. 1). Modeling operations are then defined as graph transformation rules, using two types of dedicated variables: orbit variables [19] and node variables [4] which respectively abstract topological structures (*e.g.* in order to subdivide a face whatever its number of vertices) and embeddings (*e.g.* in order to compute the barycenter of a face whatever its vertices' positions). The rule editor (see Fig. 1(a)) includes a syntactic analyzer which guides the user while ensuring consistency. Once designed, a modeler can be generated and used right away with the provided generic viewer (see Fig. 1(b)) or integrated into larger tools. End-users of this modeler are not required to understand the rule language as they only pick and apply operations interactively.

Building Consistent Objects. Jerboa's rule language relies on two key aspects: the instantiation of variables and the syntactic conditions of consistency preservation (*i.e.* conditions on rules that preserve G-map labeling constraints). As long as only one variable type is concerned, these aspects have been proved well-founded. Purely topological operations defined with the sole use of orbit variables (*e.g.* unsew a vertex, sew two faces, etc.) have been discussed in [5,19], whereas geometric operations defined with the sole use of node variables (*e.g.* translate a vertex, swap two face colors, etc.) have been discussed in [3,4]. Therefore, the previous limitation of Jerboa regarded the simultaneous use of both variable types which come with different instantiation mechanisms and syntactic conditions, making them hard to use together.

Contribution. Following the general approach of DPO graph transformations with variables defined in [11], we propose a two-layered instantiation of variables. Orbit variables are first substituted by cells of the object, thus automatically duplicating the node variables. These can then be substituted if a match morphism exists, thus leading to a classical DPO application of the instantiated rule. To ensure that transformed objects are consistent, we extend the syntactic conditions on rules separately defined for each variable type to rules with both variable types.

Related Work. Formal rule languages are already commonly used in the context of geometric modeling. In particular, L-systems are particularly useful to procedurally model regular objects such as plants or buildings [6,15]. However, they are inadequate to design a generic geometric modeler since every new construction requires dedicated implementation efforts. Conversely, graph transformation rules are self-contained and can be applied with a single rule application engine (for a given transformation class). Moreover, they have already been enriched with several variable types with various genericity purposes (*e.g.* label computations [13], labeling constraints [16,17], structural transformations [10,12]), which facilitated the definition of variables dedicated to geometric modeling. At last, let us point out that despite the existence of many efficient generic tools (*i.e.* GrGen.NET, Groove, AGG, etc.) [14,18], we favored the development of a dedicated tool [1] for performance issues. Indeed, as any geometric modeler, modelers designed with the help of Jerboa have to interactively handle objects that can be over a million nodes large, whereas the mentioned tools only allow few thousand nodes.

Paper Organization. Section 2 presents G-maps [7] and conditions under which rules without variable define consistent transformations. Section 3 presents orbit variables and node variables and their respective conditions of consistency preservation. In Sect. 4, we introduce a dedicated two-layered instantiation process to simultaneously use both variable types and we provide new conditions for consistency preservation.

2   G-maps and Their Transformations

2.1   G-maps

The topological model of G-maps [7] can be directly encoded with labeled graphs: the topological structure is defined by both the graph itself and the arc labels, while the embedding is defined by node labels. To handle multiple embeddings (*e.g.* vertex positions and face colors in Fig. 2(a)), we defined in [3] the category of Π-*graphs* (with Π a finite set of node labels) as an extension of partially labeled graphs [9], in which nodes have $|\Pi|$ labels.

In the sequel, $n \in \mathbb{N}$ will denote the dimension of considered objects and τ will denote a generic data type name with $\lfloor \tau \rfloor$ its set of typed values.

Π-**graph.** For $\Pi = (\pi :\to \tau)$ a family of typed names, a Π-*graph* $G = (V, E, s, t, (\pi)_{\pi \in \Pi}, \alpha)$ consists in a set E of arcs, two functions source $s : E \to V$ and target $t : E \to V$, a family of partial functions[2] $(\pi : V \to \lfloor \tau \rfloor)_{\pi \in \Pi}$ that label nodes and a partial function $\alpha : E \to [0, n]$ that labels arcs.

---

[2] Given X and Y two sets, a partial function f from X to Y is a total function $f : X' \to Y$, from X' a subset of X. X' is called the domain of f, and is denoted by $Dom(f)$. For $x \in X \backslash Dom(f)$, we say that $f(x)$ is undefined, and write $f(x) = \bot$. We also note $\bot : X \to Y$ the function totally undefined, that is $Dom(\bot) = \emptyset$.

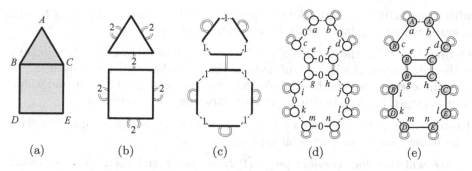

Fig. 2. Decomposition of a geometric 2D object into a 2-G-map (Color figure online)

All examples will be colored G-maps in dimension 2, such as in Fig. 2. Arcs are labeled on $[0, 2]$ to encode topological relations, while nodes are labeled by positions and colors (functions *pos* and *col* respectively) to encode embedding data ($\lfloor\tau_{\text{pos}}\rfloor = [A, B, C, D, \ldots]$, $\lfloor\tau_{\text{col}}\rfloor = [\bullet, \circ, \bullet, \circ, \ldots]$).

G-maps intuitively result from the object decomposition into topological cells. The 2D object of Fig. 2(a) is first (Fig. 2(b)) decomposed into faces connected along their common edge with a 2-relation and provided with 2-loops on border edges. Similarly, faces are split into edges connected with 1-relations (Fig. 2(c)). At last, edges are split into vertices by 0-relations to obtain the 2-G-map of Fig. 2(d). Nodes obtained at the end of the process are the G-map nodes and the different i-relations are labeled arcs: for a G-map of dimension n, i belongs to $[0, n]$. For readability purpose, we will use the graphical codes of Fig. 2 (black line for 0-arcs, red dashed line for 1-arcs and blue double line for 2-arcs).

Topological cells of G-maps are defined by subgraphs called *orbits* and built from an originating node v and a set $o \subseteq [0, n]$. By denoting $\langle o \rangle$ any word on o without repetition (e.g. $\langle 1\,2 \rangle$ or $\langle 2\,1 \rangle$ for $o = \{1, 2\}$), the orbit $\langle o \rangle(v)$ (of type $\langle o \rangle$ adjacent to v) is the subgraph which contains v, the nodes reachable from node v using arcs labeled on o, and the arcs themselves. By definition, embedding data (positions or colors) are shared by all nodes belonging to an orbit of the associated type. In Fig. 2(e), the vertex adjacent to e is the orbit $\langle 1\,2 \rangle(e)$ which contains nodes c, e, g and i, all labeled with the position B attached to the vertex. Similarly, nodes a, b, c, d, e, f all belong to the same face orbit $\langle 0\,1 \rangle$ and are all labeled by the same color $\bullet$.

Topological graph. An Π-graph G is a n-*topological graph* if the arc labeling function α is a total function on $[0, n]$. G_α, called the topological structure of G, is the graph G, except that all node labeling functions are totally undefined.

Orbit type. An *orbit type* $\langle o \rangle$ is a subset $o \subseteq [0, n]$, and denoted as a word on $[0, n]$ without repetition.

Orbit equivalence. For any orbit type $\langle o \rangle$, $\equiv_{G\langle o \rangle}$ is the *orbit equivalence relation* defined on $V_G \times V_G$ as the reflexive, symmetric and transitive closure built from arcs

with labels in o (*i.e.* ensuring for each arc $e \in G$ with $\alpha(e) \in o$, $s_G(e) \equiv_{G\langle o \rangle} t_G(e)$).
A graph whose edge labels are in $\langle o \rangle$ is said to be of type $\langle o \rangle$.

Orbit. For $v \in V_G$, the $\langle o \rangle$-*orbit* of G adjacent to v, denoted by $G\langle o \rangle(v)$ (or $\langle o \rangle(v)$),
is the subgraph of G whose set of nodes is the equivalence class of v using $\equiv_{G\langle o \rangle}$
and whose set of arcs are those labeled on o between nodes of $G\langle o \rangle(v)$, and whose
source, target, labeling functions are the restrictions of functions of G.

Embedding. An *embedding* $\pi : \langle o \rangle \rightarrow \tau$ is characterized by a name π, a data
type name τ and a support orbit type $\langle o \rangle$.

We will therefore consider $pos : \langle 1\ 2 \rangle \rightarrow point$ and $col : \langle 0\ 1 \rangle \rightarrow color$.
Regarding an embedding $\pi : \langle o \rangle \rightarrow \tau$, all nodes of an $\langle o \rangle$-orbit will share the same
label by π (called π-label). G-maps are provided with an embedding constraint
capturing this property [3]. Besides, G-maps are equipped with constraints relating to the topology. The cycle constraint ensures that in G-maps, two i-cells can
only be adjacent along $(i-1)$-cells. For instance, in Fig. 2(d), the 0202-cycle
constraint implies that faces are stuck along topological edges.

Definition 1 (G-map [3,7,19]). *For $\Pi = (\pi : \langle o \rangle \rightarrow \tau)$ a family of embeddings, a G-map embedded on Π (or Π-embedded G-map) is a topological graph
$G = (V, E, s, t, (\pi)_{\pi \in \Pi}, \alpha)$ that satisfies the following consistency constraints:*

- *Symmetry: G is symmetric (i.e. $\forall e \in E$, $\exists e' \in E$, such that $s(e') = t(e)$,
  $t(e') = s(e)$, and $\alpha(e') = \alpha(e)$),*
- *Adjacent arcs: each node is the source node of $n+1$ arcs labeled from 0 to n,*
- *Cycles: $\forall i, j$ such that $0 \leq i \leq i+2 \leq j \leq n$, there exists a cycle[3] labeled by
  $ijij$ starting from each node.*
- *Embedding consistency: every node labeling function $\pi \in \Pi$ is total and for
  all nodes v and w such that $v \equiv_{\langle o \rangle} w$, $\pi(v) = \pi(w)$.*

2.2 Consistent G-map Transformations Using DPO

Morphisms on Π-graphs extend morphisms defined on partially labeled graphs
to a set of labels Π. In [3], we extended relabeling graph transformations of [9]
to Π-graphs using the double-pushout approach (DPO) [8].

Morphism. For two Π-graphs $G = (V_G, E_G, s_G, t_G, (\pi_G)_{\pi \in \Pi}, \alpha_G)$ and $H = (V_H, E_H, s_H, t_H, (\pi_H)_{\pi \in \Pi}, \alpha_H)$, a Π-*graph morphism* $m : G \rightarrow H$ is defined by
two functions $m_V : V_G \rightarrow V_H$ and $m_E : E_G \rightarrow E_H$ preserving sources, targets
and labels - *i.e.* $s_H \circ m_E = m_V \circ s_G$, $t_H \circ m_E = m_V \circ t_G$, $\pi_H(m(v)) = \pi_G(v)$
for all $\pi \in \Pi$ and $v \in Dom(\pi_G)$, and $\alpha_H(m(e)) = \alpha_G(e)$ for all $e \in Dom(\alpha_G)$.
If $m(v) = v$ for all $v \in V_G$ and $m(e) = e$ for all $e \in E_G$, m is an *inclusion* and is
denoted $m : G \hookrightarrow H$.

We have shown in [3] that the Π-graph category inherits from the partially labeled graph category [9] all classical properties such as the existence
of pushouts.

---

[3] A cycle is a sequence $e_1...e_k$ of arcs such $t(e_i) = s(e_{i+1})$ for each $1 \leq i < k$ and
$t(e_k) = s(e_1)$. The word $\alpha(e_1)...\alpha(e_k)$ is called its label.

Rule. A rule $r : L \hookleftarrow K \hookrightarrow R$ consists of two inclusions $K \hookrightarrow L$ and $K \hookrightarrow R$ such that:

- for all $v \in V_L$ (resp. for all $e \in E_L$ and all $\pi \in \Pi$), $\alpha_L(v) = \bot$ (resp. $\pi_L(e) = \bot$) implies $v \in V_K$ and $\alpha_R(v) = \bot$ (resp. $e \in E_K$ and $\pi_R(e) = \bot$),
- for all $v \in V_R$ (resp. for all $e \in E_R$ and all $\pi \in \Pi$), $\alpha_R(v) = \bot$ (resp. $\pi_R(e) = \bot$) implies $v \in V_K$ and $\alpha_L(v) = \bot$ (resp. $e \in E_K$ and $\pi_L(e) = \bot$).

We call L the *left-hand side*, R the *right-hand side* and K the *interface* of r.

Direct derivation. A *direct derivation* from a graph G to a graph H *via* a rule $r : L \hookleftarrow K \hookrightarrow R$ consists of two natural pushouts [9] as in Fig. 3, where $m : L \to G$, called a *match morphism*, is injective[4]. If this derivation exists, we write $G \Rightarrow_{r,m} H$.

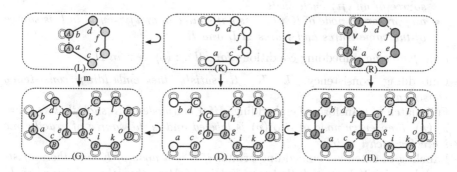

Fig. 3. A direct derivation (Color figure online)

For example, the rule $L \hookleftarrow K \hookrightarrow R$ in Fig. 3 matches a blue triangle which has a vertex at position A. The rule transforms the triangle into a square by splitting this vertex into two vertices at positions I and J, while changing the face color to red. According to the rule definition, unmatched labels in L such as the position of node d are also unmatched in R. Secondly, nodes of R with undefined labels are preserved nodes of the rule, with undefined labels in L. Consequently, added nodes of R such as the node v have all their labels defined in R (I and ●). Moreover, changed labels such as the color of d in R (●) are matched in L (●). These rule properties ensure that when the rule is applied to a totally labeled object such as G in Fig. 3, the resulting graph H is also totally labeled.

To ensure that direct derivations preserve G-map constraints of Definition 1, [19] and [3] respectively introduced *topological conditions* and *embedding conditions* on rules that ensure consistency preservation. For example, the rule of Fig. 3 preserves consistency as the added nodes u and v are added with all their adjacent arcs, cycles and embeddings. Similarly, the embedding modifications (● to ● and A to I/J) are consistently defined on all concerned nodes.

---

[4] A morphism g is injective if g_V and g_E are injective.

Result 1 (G-map consistency preservation using basic rules [3,4,19]**).**
For $r : L \hookleftarrow K \hookrightarrow R$ a graph transformation rule and $m : L \to G$ a match morphism on a Π-embedded n-G-map, the direct transformation $G \Rightarrow^{r,m} H$ produces a Π-embedded n-G-map if r satisfies the following topological conditions:

- Symmetry: *L, K and R satisfy the symmetry constraint.*
- Adjacent arcs:
  - *preserved nodes of K are sources of arcs having the same labels in both the left-hand side L and the right-hand side R;*
  - *removed nodes of $L \backslash K$ and added nodes of $R \backslash K$ must be source of exactly $n + 1$ arcs respectively labeled from 0 to n.*
- Cycles: *for all pair (i, j) such $0 \leq i \leq i + 2 \leq j \leq n$,*
  - *any added node of $R \backslash K$ is source of an $ijij$-cycle;*
  - *any preserved node of K which is source of an $ijij$-cycle in L, is also source of an $ijij$-cycle in R;*
  - *any preserved node of K which is not source of an $ijij$-cycle in L is source of the same i-arcs and j-arcs in L and R.*

and the following embedding conditions for all $\pi : \langle o \rangle \to \tau$ in Π:

- Embedding consistency: *L, K, R satisfy the embedding consistency constraint.*
- Full match of transformed embeddings: *if v is a node of K such that $\pi_L(v) \neq \pi_R(v)$, then every node of $R\langle o \rangle(v)$ is labeled and is the source of exactly one i-arc for each i of $\langle o \rangle$.*
- Labeling of extended embedding orbits: *if v is a node of K and there exists a node w in $R\langle o \rangle(v)$ such that w is not in $L\langle o \rangle(v)$, then there exists v' in K with $v' \equiv_{L\langle o \rangle} v$ and $v' \equiv_{R\langle o \rangle} v$ such that $\pi_L(v') \neq \bot$.*

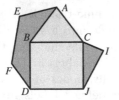   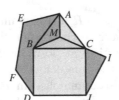   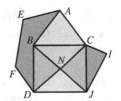

(a) Modeled object   (b) Triangle triangulation   (c) Square triangulation

Fig. 4. A modeling operation: face triangulation

3   Rule Variables for Geometric Modeling

Let us consider the face triangulation operation of Fig. 4. On the topological side, the face is subdivided into as many triangles as it contains edges. On the embedding side, new face colors are computed as the mix of the subdivided face color and the neighboring face color, while the position of the created vertex is set as the barycenter of the transformed face. As the operation depends on the attributes (number of vertices, position, color) of the matched object, its definition by a single generic rule requires the use of variables.

3.1   Graph Transformations with Variables

Intuitively, *rule schemes* (rules with variables) describe as many concrete rules as there are possibilities to instantiate variables with concrete elements. In [11], a generic approach has been proposed to deal with variables within graph transformations: roughly speaking, a rule scheme is first applied to a graph along a *kernel match morphism* (in our case a morphism that only matches the topological structure L_α of the left-hand side L). Then, if a substitution σ associating a value for all variables can be induced from the kernel match morphism, an instance rule, generically denoted $L^\sigma \hookleftarrow K^\sigma \hookrightarrow R^\sigma$ for a rule scheme $L \hookleftarrow K \hookrightarrow R$, can be built and finally, applied to the graph as a classical rule.

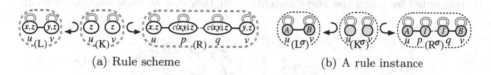

(a) Rule scheme   (b) A rule instance

Fig. 5. Embedding label computations with attribute variables

Using this principle, [11] introduces three types of variables. Let us briefly present two of them which relate to our dedicated variable types. First, *attribute variables* illustrated in Fig. 5 allow label computation. In the rule scheme of edge subdivision, variables x and y abstract two position labels while z abstracts a color one. In the right-hand side, a topological vertex is added at the center of the existing two positions, using the dedicated center operator $c : point \times point \rightarrow point$, while the color is set to the same face color z.

Second, *clone variables* illustrated in Fig. 6 allow structural abstraction. Intuitively, in order to subdivide n edges, nodes of the rule scheme which are labeled by the clone variable n are duplicated by as many nodes as the multiplicity value used to instantiate n, while arcs are duplicated accordingly to node duplications and arcs of the rule scheme. Note that as we also use the attribute variables x, y, and z, the rule scheme of Fig. 6(b) still requires to substitute them.

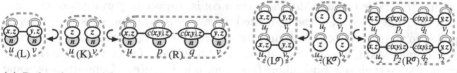

(a) Rule scheme with n as clone variable   (b) Instantiation of n by 2

Fig. 6. Expanding computations with clone variables

3.2   Node Variables for Embedding Computation

Embedding computations require to traverse the topological structure of modified objects: *e.g.* the face triangulation of Fig. 4 requires to access the colors of adjacent faces, whether such faces exist or not. In [4], we therefore introduced *node variables* which are similar to attribute variables. By directly using node names of L as variables, we provide operators on node variables to access embedding labels and adjacent nodes, thanks to G-map regularity. For a node variable a, $a.\pi$ gives access to its π-label while $a.\alpha_i$ (with $i \leq n$) gives access to the node connected to node a by an i-arc. Embedding expressions used in a rule scheme $L \hookleftarrow K \hookrightarrow R$ are then terms built over these operators and nodes of L. Thus, for a kernel match morphism $m : L_\alpha \to G$, a rule instance $L^{mv} \hookleftarrow K^{mv} \hookrightarrow R^{mv}$ can be computed by using the node matching function m_V as variable substitution.

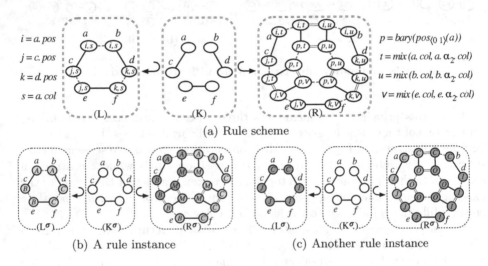

(a) Rule scheme

(b) A rule instance   (c) Another rule instance

Fig. 7. Triangle triangulation with embedding terms

From the rule scheme of Fig. 7(a), we can build the two rule instances of Fig. 7(b) and (c) respectively corresponding to triangles BAC and JCI of Fig. 4(a). In the left-hand side, the term $a.col$ is respectively evaluated as ● and ●. In the right-hand side, the dedicated operator $mix : color \times color \to color$ is applied to the face color ($a.col$) and the color of the neighboring face ($a.\alpha_2.col$). At rule application, $a.\alpha_2$ is evaluated as the 2-neighbor of a. Note that in the border, nodes are their self 2-neighbors (*e.g.* node a in Fig. 2(e)). Consequently, some created faces in Fig. 7(b) and (c) keep their original color. At last, embedding expressions also include operators in charge of collecting all embedding values carried by a given orbit. In the rule scheme of Fig. 7(a), the term $pos_{\langle 0\ 1 \rangle}(a)$ is evaluated as the multiset of positions labeling the face, i.e. of $\langle 0\ 1 \rangle$-orbit adjacent to node a. In the instance of Fig. 7(b), this set is $[A, B, C]$ and the added vertex is positioned at $bary([A, B, C]) = M$ using a dedicated barycenter operator.

In order to instantiate a rule scheme built over node variables into a rule that satisfies the conditions of Result 1, we introduced in [4] a completion step. Let us consider the example of Fig. 8 that merges two faces by removing their common edge and mixing their colors. This operation can be defined independently of face shapes by the minimal rule scheme of Fig. 8(a) that only deals with the central edge and the adjacent nodes. The completion step automatically includes in the rule instance all nodes concerned by embedding modifications. In the example of Fig. 8(b) corresponding to an application to the object of Fig. 2(e), this step completes the rule with the rest of the two faces so that the green color can be attached on the whole new face.

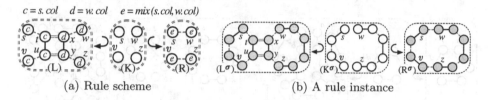

(a) Rule scheme   (b) A rule instance

Fig. 8. Face merge with embedding terms (Color figure online)

Consequently, rule schemes are exempt from the condition of full match of transformed embeddings of Result 1. However, to prevent misapplication cases in which two different embedding orbits would be matched as a single one, the condition of full match is replaced by a non-overlap condition on match morphism. Note that similarly to the condition of injective match morphism, the non-overlap condition has to be dynamically checked, with no particular difficulty.

Result 2 (G-map consistency preservation using node variables [4]).
For $r : L \hookleftarrow K \hookrightarrow R$ a rule scheme with node variables and $m : L_\alpha \to G$ a match morphism on a Π-embedded G-map, if r satisfies the conditions of Result 1, except the full match of transformed embeddings, and satisfies the following non-overlap condition for all $\langle o \rangle$ occurring as support orbit type in $\Pi = (\pi : \langle o \rangle \to \tau)$, then the instance rule[5] $r^{mv} : L^{mv} \hookleftarrow K^{mv} \hookrightarrow R^{mv}$ satisfies the conditions of Result 1.
Non-overlap: *for $v, u \in V_L$ such as $v \not\equiv_{L\langle o \rangle} u$, $m(v) \not\equiv_{G\langle o \rangle} m(u)$.*

3.3 Orbit Variables for Topological Rewriting

As existing variable types were unfit to abstract G-map cell transformations, we introduced *orbit variables* in [19]. Intuitively, these are typed by an orbit type $\langle o \rangle$ so that they can abstract any G-map orbit of type $\langle o \rangle$. By rewriting $\langle o \rangle$

---

[5] Note that the instantiation includes the completion step that consist in extending the matched and transformed patterns to the embedding orbits [4].

into another type $\langle o' \rangle$, we can change arc labels or remove arcs. For example, the rule scheme of Fig. 9(a) models the topological triangulation of any face - *i.e.* any orbit of type $\langle 0\ 1 \rangle$. Note that, as for clone variables in Fig. 6, the other node labels, in this case the color labels, are duplicated along the orbit variable instantiation. Note also that these colors have been chosen to help reading of orbit copies, disregarding G-map embedding consistency preservation.

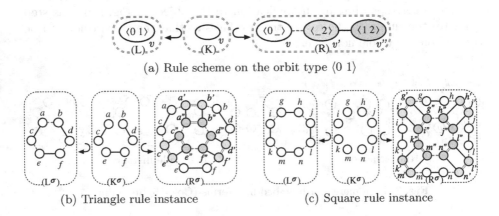

(a) Rule scheme on the orbit type $\langle 0\ 1 \rangle$

(b) Triangle rule instance   (c) Square rule instance

Fig. 9. Face triangulation with orbit variable

The two instance rules of Fig. 9(b) and (c) are constructed as follows:

1. the node v labeled in L with the orbit type $\langle 0\ 1 \rangle$ is substituted by a face, i.e. a $\langle 0\ 1 \rangle$-orbit, e.g. a triangle (resp. a square), to build L^σ ;
2. each node of L^σ is kept in R^σ, and duplicated twice, corresponding to the nodes v, v' and v'', labeled by different orbit types in R;
3. for the face matched by node v, 0-arcs are conserved in R^σ while 1-arcs are removed as v is relabeled in R by $\langle 0\ \_ \rangle$ (an empty label "$\_$" replaces 1); similarly, for node v', the type $\langle \_\ 2 \rangle$ means both removal of 0-arcs and 2-relabeling of 1-arcs while for the v'' node, $\langle 1\ 2 \rangle$ entails 1-relabeling of 0-arcs and 2-relabeling of 1-arcs;
4. at last, as indicated by the arcs of R between v, v' and v'' nodes, any node of the matched face v is connected to its image in copy v' with a 1-arc, and all v' and v'' images of a given node are connected with a 0-arc.

Topological rewriting. A *topological rewriting* $\langle \omega \rangle$ of an orbit type $\langle o \rangle$ is defined by a word ω on $[0,n] \cup \{\_\}$, of same length as o, and such that for all i in $[0,n]$, there is at most one occurrence of i in ω. We write $\langle o \rightarrow \omega \rangle$ for the *type rewriting function* that associates each label $o_i \in [0,n]$ at position i in $\langle o \rangle$ to its images ω_i at the same position in $\langle \omega \rangle$ - *i.e.* $\langle o \rightarrow \omega \rangle(o_i) = \omega_i$. For a graph $G = (V, E, s, t, (\pi)_{\pi \in \Pi}, \alpha)$ of type $\langle o \rangle$, we denote $G_{\langle o \rightarrow \omega \rangle} = (V, E', s, t, (\pi)_{\pi \in \Pi}, \alpha')$ the rewritten graph with $E' \subset E$ such that $\forall e \in E$, $e \notin E'$ if $\langle o \rightarrow \omega \rangle(\alpha(e)) = \_$, and $e \in E'$ otherwise with $\alpha'(e) = \langle o \rightarrow \omega \rangle(\alpha(e))$. At last, for convenience, $\langle o \rangle$ denotes both an orbit type and the identity type rewriting function $\langle o \rightarrow o \rangle$.

Rule scheme with orbit variable. For an orbit type $\langle o \rangle$, a *rule scheme* $L \hookleftarrow K \hookrightarrow R$ *with orbit variable* $\langle o \rangle$ is such that all nodes of L, K and R are labeled by topological rewritings of $\langle o \rangle$ and at least one node of L is labeled by $\langle o \rangle$.

Orbit variable instantiation. For $r : L \hookleftarrow K \hookrightarrow R$ a rule scheme with orbit variable $\langle o \rangle$ and O a graph of type $\langle o \rangle$, we denote $r^O : L^O \hookleftarrow K^O \hookrightarrow R^O$ the instantiated rule[6] [19]. The functions that respectively associate instance nodes with their originating nodes in graphs of r or in O are respectively denoted $\uparrow_{L^O}^L : V_{L^O} \rightarrow V_L$, $\uparrow_{K^O}^L : V_{K^O} \rightarrow V_K$, $\uparrow_{R^O}^R : V_{R^O} \rightarrow V_R$ and $\uparrow_{r^O}^O : (V_{L^O} \cup V_{K^O} \cup V_{R^O}) \rightarrow V_O$. In particular, for all $\pi \in \Pi$ and all node v of r^O, we have $\pi(v) = \pi(\uparrow_{r^O}^O (v))$.

Since nodes of rule schemes contain topological rewritings as special labels used to match topological graphs, they do not belong to the same Π-category than the underlying Π-embedded G-maps. However, as in Fig. 9, these extra labels disappear after variable instantiation. Additionally, the syntactic conditions that preserve G-map topological consistency have been adapted to handle both the explicit arcs of rule graphs and the implicit arcs of topological rewritings that label nodes. For example, node v' in Fig. 9(a) is added with all its adjacent arcs as both the 0-arc and 1-arc are explicit in R, while the 2-arc is implicit in the orbit rewriting $\langle \_\ 2 \rangle$. Similarly, v' is added with an half-implicit 0202-cycle as v' and v'' are connected with an explicit 0-arc while 2 is at the same position in their topological rewritings $\langle \_\ 2 \rangle$ and $\langle 1\ 2 \rangle$.

Result 3 (Topological consistency preservation using orbit variable [19]**).** *For* $r : L \hookleftarrow K \hookrightarrow R$ *a rule scheme with orbit variable* $\langle o \rangle$ *and a graph* O *of type* $\langle o \rangle$, *the instance rule* $r^O : L^O \hookleftarrow K^O \hookrightarrow R^O$ *satisfies the topological conditions of G-map consistency preservation of Result 1 if* r *satisfies the same conditions extended to implicit arcs and cycles such as for* v *a node of* L, K *or* R *labeled by* $\langle \omega \rangle$:

- *for* $i \in [0, n]$, v *is source of an* implicit i-arc *if* $i \in \langle \omega \rangle$; *for* $v \in K$, *this implicit* i-arc *is preserved if* i *is at the same position in* $\langle \omega_L \rangle$ *and* $\langle \omega_R \rangle$ *the respective labels of* v *in* L *and* R - *i.e.* $\langle \omega_L \rangle \rightarrow \langle \omega_R \rangle(i) = i$;
- *for all* (i, j), v *is source of an* implicit $ijij$-cycle *if* $i \in \langle \omega \rangle$, $j \in \langle \omega \rangle$, *and there exists a node* v' *in* L *labeled by* $\langle \omega' \rangle$ *and source of an* $i'j'i'j'$-cycle *such that* $i' = \langle \omega \rangle \rightarrow \langle \omega' \rangle(i)$ *and* $j' = \langle \omega \rangle \rightarrow \langle \omega' \rangle(j)$;
- *for all* (i, j), v *is source of an* half-implicit $ijij$-cycle *if* $i \in \langle \omega \rangle$ *(resp.* $j \in \langle \omega \rangle$*)* *and either* v *is source of an* j-loop *(resp.* i-loop*) or* v *is connected by an* j-arc *(resp.* i-arc*) to a node* v' *labeled by* $\langle \omega' \rangle$ *such that* i *(resp.* j*) is at the same position in* $\langle \omega \rangle$ *and* $\langle \omega' \rangle$ - *i.e.* $\langle \omega \rangle \rightarrow \langle \omega' \rangle(i) = i$ *(resp.* $\langle \omega \rangle \rightarrow \langle \omega' \rangle(j) = j$*)*.

As orbit variables abstract multiple nodes (and arcs), their combined use with node variable requires some care in order to instantiate both variable types.

---

[6] L^O, K^O, and R^O are the Cartesian product graphs of O and resp. graphs L, K and R, by keeping tracks of arc relabelings and arc removals.

4   Rule Schemes for Specifying Modeling Operations

4.1   Combining Orbit Variables and Node Variables

Let us consider the rule scheme of Fig. 10(a) that defines the face triangulation of Fig. 7 by combining the two variable transformations of Figs. 7 and 9. Intuitively, as the orbit variable $\langle 0\ 1\rangle$ defines the topological structure of the rule instance, it has to be substituted first in order to provide all the node variables required to match the embedding of the face. When the orbit variable is instantiated in Fig. 10(b) by a triangle face, the terms $v.pos$ and $v.col$ are duplicated on all instantiated nodes. However, they are rewritten by respectively replacing v by the new node names a, b, $\ldots$, f. For example, the term $mix(v.col, v.\alpha_2.col)$ has been rewritten on c'' by $mix(c.col, c.\alpha_2.col)$ as c is the corresponding new variable.

Definition 2 (Term rewriting). *Let* $r : L \hookleftarrow K \hookrightarrow R$ *be a rule scheme with orbit variable* $\langle o \rangle$ *and node variables. Let* r^O *be a rule scheme with node variables resulting from a substitution* $\langle o \rangle$ *by a graph* O *of type* $\langle o \rangle$.

The rewritten rule scheme $r(O)$ *results from the respective application of the following term functions* ρ_v *for each node* v *of* r^O *to the node labels of* r^O, *such that* ρ_v *extends the following variable substitution: for every node variable* u *of* V_L, $\rho_v(u) = u'$ *in which* u' *is the unique node variable of* V_{L^O} *such that* $\uparrow_{L^O}^{L}(u') = u$ *and* $\uparrow_{r^O}^{O}(u') = \uparrow_{r^O}^{O}(v)$.

Similarly to the rule scheme of Fig. 7(a), the rewritten one of Fig. 10(a) only requires to substitute node variables to produce the instance rules of Fig. 7(b) and (c), but in this case multiple terms define the same value. For example, the barycenter is successively defined as $bary(pos_{\langle 0\ 1\rangle}(a))$, $bary(pos_{\langle 0\ 1\rangle}(b))$, $\ldots$, $bary(pos_{\langle 0\ 1\rangle}(f))$ which will all result in the same position. Therefore we must adapt the conditions of consistency preservation.

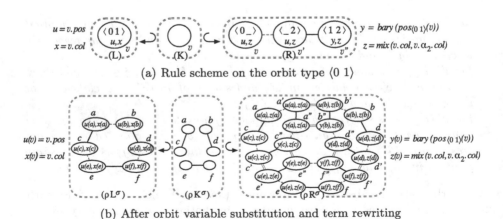

(a) Rule scheme on the orbit type $\langle 0\ 1\rangle$

(b) After orbit variable substitution and term rewriting

Fig. 10. Face triangulation with both variables types

4.2   Consistency Preservation

Let us consider the rule scheme of Fig. 11(a) that still defines the triangulation, but with different embedding computations. In particular, the center is positioned right between the corner and the barycenter $(c(v.pos, bary_{\langle 0\ 1\rangle}(v)))$ and the color of created faces is defined as the mix between the original color and the color of the adjacent face around the corner $(v.\alpha_1.\alpha_2.col)$. Note that this rule satisfies the conditions of embedding consistency preservation as defined in Result 2. However, most instances of this scheme surely break the embedding consistency. Figure 11(b) and (c) respectively present the rule instance and the intuitive representation for its application to the triangle ABC of Fig. 4(a). This rule breaks the added vertex consistency since three different positions are computed: $I = c(A, M)$, $J = c(B, M)$, $K = c(C, M)$ with $M = bary([A, B, C])$. Similarly, each created face is embedded with two different colors since the original color is mixed with the colors of the faces around the two corners.

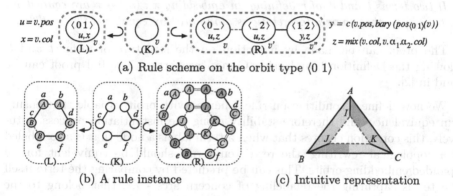

(a) Rule scheme on the orbit type $\langle 0\ 1 \rangle$

(b) A rule instance   (c) Intuitive representation

Fig. 11. Break of G-map consistency

To prevent inconsistencies, conditions on rule scheme must take into account the pattern expanding and the term rewriting due to the orbit variable substitution. For this purpose, considering rule schemes without orbit variable, we first define the equivalence between two embedding terms that ensures embedding equality on rule application to a G-map.

Definition 3 (Term equivalence). *Let $r : L \hookleftarrow K \hookrightarrow R$ be a rule scheme with node variables.*

For any terms t and t' of r that define an embedding $\pi : \langle o \rangle \to \tau$, the equivalence between terms, denoted $t \equiv_{L\langle o \rangle} t'$, is the smallest equivalence relation extending the equivalence between nodes of L such that:

- *for all dimension $i \in [0, n]$, let $\langle o' \rangle$ the sub-orbit of $\langle o \rangle$ containing all dimensions j of $\langle o \rangle$ such $j + 2 \le i$ or $i + 2 \le j$, if $t \equiv_{L\langle o' \rangle} t'$ then $t.\alpha_i \equiv_{L\langle o' \rangle} t'.\alpha_i$,*
- *for all dimension $i \in \langle o \rangle$, if $t \equiv_{L\langle o \rangle} t'$ then $t \equiv_{L\langle o \rangle} t'.\alpha_i$ and $t.\alpha_i \equiv_{L\langle o \rangle} t'$,*

– *for all embedding* $\pi' : \langle o' \rangle \to \tau'$, *if* $t \equiv_{L\langle o' \rangle} t'$ *then* $t.\pi' \equiv_{L\langle o \rangle} t'.\pi'$,
– *for all orbit* $\langle o' \rangle$, *if* $t \equiv_{L\langle o' \rangle} t'$ *then* $\pi_{\langle o' \rangle}(t) \equiv_{L\langle o \rangle} \pi_{\langle o' \rangle}(t')$,
– *for all user function* $f : s_1 \times ... \times s_m \to s_{m+1}$ *and all terms* $t_1, t_1' : s_1, ..., t_m, t_m' :$
 s_m, *if* $t_1 \equiv_{L\langle o \rangle} t_1'$ *and* ... *and* $t_m \equiv_{L\langle o \rangle} t_m'$ *then* $f(t_1, ..., t_m) \equiv_{L\langle o \rangle} f(t_1', ..., t_m')$.

Intuitively, in the rule scheme of Fig. 10(b), the two terms $bary(pos_{\langle 0\ 1 \rangle}(a))$ $bary(pos_{\langle 0\ 1 \rangle}(b))$ are equivalent because a and b belong to the same $\langle 0\ 1 \rangle$-orbit in L. Similarly, $mix(a.col, a.\alpha_2.col)$ and $mix(c.col, c.\alpha_2.col)$ are equivalent because: (i) $a.col$ and $c.col$ are equivalent as they belong to the same $\langle 0\ 1 \rangle$-orbit carrying the color embedding; (ii) $a.\alpha_2.col$ and $c.\alpha_2.col$ are equivalent as the α_2-access ensures that $a.\alpha_2$ and $c.\alpha_2$ belong to the same $\langle 0\ 1 \rangle$-orbit and thus have the same color.

Theorem 1 (Evaluation equality along term equivalence). *Let* $r : L \hookleftarrow$ $K \hookrightarrow R$ *be a rule scheme with node variables and* $m : L_\alpha \to G$ *a kernel match morphism on a Π-embedded n-G-map G.*

If two terms t and t' of r defining an embedding $\pi : \langle o \rangle \to \tau$ are equivalent, i.e. $t \equiv_{L\langle o \rangle} t'$, then their interpretations along m_V are equal, i.e. $t^{m_V} = t'^{m_V}$.

The proof can be made by induction on the structure of terms t and t' following the Definition 3, and using G-map properties. The full proof can be found in [2].

We now define a condition on rule schemes with orbit variable that ensure term equivalence, and therefore stability along the instantiation process. Intuitively, this condition ensures that when a term labels a node that is also labeled by a topological rewriting, the rewritten terms should be equivalent for all expanded embedding orbits. This can be predicted by comparing the term itself with terms capturing the relabeling of concern arcs - *i.e.* that belong to the embedding orbit.

Let us consider the term $bary(pos_{\langle 0\ 1 \rangle}(v))$ that labels node v'' in the rule scheme of Fig. 10(a). As node v'' is labeled by $\langle 1\ 2 \rangle$ and will be expanded, we must ensure equivalence for all labels of the $\langle 1\ 2 \rangle$-subset of $\langle 1\ 2 \rangle$ (vertex orbit carrying the position embedding), therefore in this case for both 1 and 2. As node v appearing in the term is originally labeled $\langle 0\ 1 \rangle$ in L, we have to consider the respective reverse relabeling $\langle 1\ 2 \to 0\ 1 \rangle(1) = 0$ and $\langle 1\ 2 \to 0\ 1 \rangle(2) = 1$. Therefore, the term must be equivalent to both $bary(pos_{\langle 0\ 1 \rangle}(v.\alpha_0))$ and $bary(pos_{\langle 0\ 1 \rangle}(v.\alpha_1))$. This is indeed true as $v.\alpha_0$ and $v.\alpha_1$ both belong to the collected orbit $\langle 0\ 1 \rangle(v)$.

Similarly, let us consider the term $mix(v.col, v.\alpha_2.col)$ labeling v''. We must ensure equivalence for all labels of the $\langle 1\ 2 \rangle$-subset of the $\langle 0\ 1 \rangle$ (face orbit carrying the color embedding), therefore in the case only 1. As node v appearing in the term is originally labeled by $\langle 0\ 1 \rangle$ in L, we consider $\langle 1\ 2 \to 0\ 1 \rangle(1) = 0$ and show that the term is equivalent to $mix(v.\alpha_0.col, v.\alpha_0.\alpha_2.col)$. As α_0 belong to the face orbit $\langle 0\ 1 \rangle$ carrying the color embedding, $v.col$ and $v.\alpha_0.col$ are equivalent. Similarly, because of the 0202-cycle constraint of G-maps, $v.\alpha_2$ and $v.\alpha_0.\alpha_2$ belong to the same face, therefore $v.\alpha_2.col$ and $v.\alpha_0.\alpha_2.col$ are equivalent terms.

Definition 4 (Condition of term stability). *Let $r : L \hookleftarrow K \hookrightarrow R$ be a rule scheme with orbit variable and node variables, and v a node of L, K or R, such v is labeled by an embedding term t defining an embedding $\pi : \langle o \rangle \to \tau$ and by a topological rewriting $\langle \omega_v \rangle$.*

The term t is stable along instantiation if for all label i of $\langle o' \rangle$ the $\langle o \rangle$-subset of $\langle \omega_v \rangle$, t is equivalent to the rewritten term t_i (i.e. $t \equiv_{L\langle o \rangle} t_i$) in which any occurence of a variable $x \in V_L$ is replaced by $x.\alpha_j$, in which $j = \langle \omega_v \to \omega_x \rangle(i)$[7] with $\langle \omega_x \rangle$ the topological rewriting that labels x in L.

Finally, rule schemes containing both node variables and orbit variables preserve G-map consistency if they satisfy all conditions previously introduced, including term stability. Note that the non-overlap condition of Result 2 is still to be additionally checked on the application morphism.

Theorem 2 (G-map consistency preservation using both variable types). *For a rule scheme $r : L \hookleftarrow K \hookrightarrow R$ with orbit variable and node variables, any rule instance resulting from the orbit variable substitution satisfies the conditions of Result 2 if r satisfies the topological conditions of Result 3 and the following embedding conditions:*

- *Embedding consistency of Result 1;*
- *Labeling of extended embedding conditions orbits of Result 1;*
- *Term stability of Definition 4.*

The proof can be made thanks to the following property resulting from the orbit variable substitution and the term rewriting: two linked nodes of a same embedded orbit come either from one node of the rule scheme r and two linked nodes of the substituted orbit O, or from two linked nodes of r and one node of O. In both cases, previous results and Theorem 1 ensure value equality. The full proof can be found in [2].

5   Conclusion

In this paper, we have presented a rule-based language for geometric modeling involving two types of variables, node variables and orbit variables, and provided them with a two-layered variable substitution mechanism. Orbit variables are first substituted, therefore defining the resulting topological structure and generating new node variables; these node variables are then substituted to compute the new embeddings. Moreover, rules written in a DPO style are provided with syntactic conditions ensuring the consistency preservation of embedded G-maps - *i.e.* by construction, transformed objects are also embedded G-maps. In particular, by introducing the conditions of term stability and terms equivalence, the syntactic conditions of embedding consistency preservation have been adapted to handle the new node variables generated by the orbit variable instantiation. This language is the core of Jerboa, a tool set for designing and generating geometric modelers.

---

[7] Note that if $\langle \omega_v \to \omega_x \rangle(i) = \_$, t_i does not exist and therefore t is not stable.

References

1. Belhaouari, H., Arnould, A., Gall, P., Bellet, T.: Jerboa: a graph transformation library for topology-based geometric modeling. In: Giese, H., König, B. (eds.) ICGT 2014. LNCS, vol. 8571, pp. 269–284. Springer, Cham (2014). doi:10.1007/978-3-319-09108-2_18
2. Bellet, T., Arnould, A., Belhaouari, H., Le Gall, P.: Geometric modeling: consistency preservation using two-layered variable substitutions (extended version). Research report (2017). https://hal.archives-ouvertes.fr/hal-01509832
3. Bellet, T., Arnould, A., Le Gall, P.: Rule-based transformations for geometric modeling. In: 6th International Workshop on Computing with Terms and Graphs (TERMGRAPH 2011), Part of ETAPS, Saarbrücken (2011)
4. Bellet, T., Arnould, A., Le Gall, P.: Constraint-preserving labeled graph transformations for topology-based geometric modeling. Research report (2017). https://hal.archives-ouvertes.fr/hal-01476860
5. Bellet, T., Poudret, M., Arnould, A., Fuchs, L., Le Gall, P.: Designing a topological modeler kernel: a rule-based approach. In: Shape Modeling International Conference (SMI), Aix-en-Provence, pp. 100–112 (2010)
6. Bohl, E., Terraz, O., Ghazanfarpour, D.: Modeling fruits and their internal structure using parametric 3Gmap L-systems. Vis. Comput. **31**(6), 747–751 (2015)
7. Damiand, G., Lienhardt, P.: Combinatorial Maps: Efficient Data Structures for Computer Graphics and Image Processing. CRC Press, Boca Raton (2014)
8. Ehrig, H., Ehrig, K., Prange, U., Taentzer, G.: Fundamentals of Algebraic Graph Transformation. Monographs on Theoretical Computer Science. Springer, Heidelberg (2006)
9. Habel, A., Plump, D.: Relabelling in graph transformation. In: Corradini, A., Ehrig, H., Kreowski, H.-J., Rozenberg, G. (eds.) ICGT 2002. LNCS, vol. 2505, pp. 135–147. Springer, Heidelberg (2002). doi:10.1007/3-540-45832-8_12
10. Habel, A., Radke, H.: Expressiveness of graph conditions with variables. Electron. Commun. EASST **30** (2010)
11. Hoffmann, B.: Graph transformation with variables. In: Kreowski, H.-J., Montanari, U., Orejas, F., Rozenberg, G., Taentzer, G. (eds.) Formal Methods in Software and Systems Modeling: Essays Dedicated to Hartmut Ehrig on the Occasion of His 60th Birthday. LNCS, vol. 3393, pp. 101–115. Springer, Heidelberg (2005). doi:10.1007/978-3-540-31847-7_6
12. Hoffmann, B.: More on graph rewriting with contextual refinement. In: Echahed, R., Habel, A., Mosbah, M. (eds.) Graph Computation Models Selected Revised Papers from GCM 2014, vol. 71. Electronic Communications of the EASST (2015)
13. Hoffmann, B., Jakumeit, E., Geiß, R.: Graph rewrite rules with structural recursion. In: Mosbah, M., Habel, A. (eds.) 2nd International Workshop on Graph Computational Models (GCM 2008), pp. 5–16 (2008)
14. Jakumeit, E., Buchwald, S., Wagelaar, D., Dan, L., Hegedüs, Á., Herrmannsdörfer, M., Horn, T., et al.: A survey and comparison of transformation tools based on the transformation tool contest. Sci. Comput. Program. **85**, 41–99 (2014)
15. Müller, P., Wonka, P., Haegler, S., Ulmer, A., Van Gool, L.: Procedural modeling of buildings. ACM Trans. Graph. (TOG) **25**, 614–623 (2006)
16. Orejas, F., Lambers, L.: Symbolic attributed graphs for attributed graph transformation. Electron. Commun. EASST **30** (2010)
17. Orejas, F., Lambers, L.: Lazy graph transformation. Fundamenta Informaticae **118**(1–2), 65–96 (2012)

18. Pérez, J., Crespo, Y., Hoffmann, B., Mens, T.: A case study to evaluate the suitability of graph transformation tools for program refactoring. Int. J. Softw. Tools Technol. Transf. **12**(3–4), 183–199 (2010)
19. Poudret, M., Arnould, A., Comet, J.-P., Gall, P.: Graph transformation for topology modelling. In: Ehrig, H., Heckel, R., Rozenberg, G., Taentzer, G. (eds.) ICGT 2008. LNCS, vol. 5214, pp. 147–161. Springer, Heidelberg (2008). doi:10.1007/978-3-540-87405-8_11

Chemical Graph Transformation
with Stereo-Information

Jakob Lykke Andersen[1,9]([✉]), Christoph Flamm[2,8], Daniel Merkle[1]([✉]),
and Peter F. Stadler[2,3,4,5,6,7]

[1] Department of Mathematics and Computer Science,
University of Southern Denmark, 5230 Odense, Denmark
{jlandersen,daniel}@imada.sdu.dk
[2] Institute for Theoretical Chemistry, University of Vienna, 1090 Wien, Austria
xtof@tbi.univie.ac.at
[3] Bioinformatics Group, Department of Computer Science,
and Interdisciplinary Center for Bioinformatics,
University of Leipzig, 04107 Leipzig, Germany
stadler@bioinf.uni-leipzig.de
[4] Max Planck Institute for Mathematics in the Sciences, 04103 Leipzig, Germany
[5] Fraunhofer Institute for Cell Therapy and Immunology, 04103 Leipzig, Germany
[6] Center for Non-coding RNA in Technology and Health, University of Copenhagen,
1870 Frederiksberg, Denmark
[7] Santa Fe Institute, 1399 Hyde Park Rd, Santa Fe, NM 87501, USA
[8] Research Network Chemistry Meets Microbiology,
University of Vienna, 1090 Wien, Austria
[9] Tokyo Institute of Technology, Earth-Life Science Institute, Tokyo 152-8550, Japan
jlandersen@elsi.jp

Abstract. Double Pushout graph transformation naturally facilitates
the modelling of chemical reactions: labelled undirected graphs model
molecules and direct derivations model chemical reactions. However, the
most straightforward modelling approach ignores the relative placement
of atoms and their neighbours in space. Stereoisomers of chemical compounds thus cannot be distinguished, even though their chemical activity may differ substantially. In this contribution we propose an extended
chemical graph transformation system with attributes that encode information about local geometry. The modelling approach is based on the so-
called "ordered list method", where an order is imposed on the set of incident edges of each vertex, and permutation groups determine equivalence
classes of orderings that correspond to the same local spatial embedding.
This method has previously been used in the context of graph transformation, but we here propose a framework that also allows for partially
specified stereoinformation. While there are several stereochemical configurations to be considered, we focus here on the tetrahedral molecular
shape, and suggest general principles for how to treat all other chemically relevant local geometries. We illustrate our framework using several
chemical examples, including the enumeration of stereoisomers of carbo-
hydrates and the stereospecific reaction for the aconitase enzyme in the
citirc acid cycle.

© Springer International Publishing AG 2017
J. de Lara and D. Plump (Eds.): ICGT 2017, LNCS 10373, pp. 54–69, 2017.
DOI: 10.1007/978-3-319-61470-0_4

Keywords: Double Pushout · Chemical graph transformation system · Stereochemistry

1 Introduction

Graph transformation systems have a long history in molecular biology [24]. Applications to chemical reaction systems have evolved from abstract artificial chemistry models such as Fontana's AlChemy [13,14] based on lambda calculus. An early attempt at more realistic modelling of chemistry with graph transformation [6] and an early perspectives article [29] proposed a variety of potential applications.

Although general graph transformation tools, such as AGG [26], have also been used to implement models of chemical systems [10], there is one crucial aspect where chemistry differs from the usual setup in the graph transformation literature. The latter focusses on rewriting a single (usually connected) graph, thus yielding a traditional formal language. Chemical reactions, in contrast, usually involve multiple molecules; chemical graph transformations therefore operate on *multisets* of graphs to produce a chemical "space" or "universe" [4], see also [17] for a similar construction in the context of DNA computing. With the software package MØD [2] we have developed a versatile suite for working with this type of transformation [5]. The packages handles composition of rules and provides a domain specific language for graph language generation [3,4].

Mathematical models for molecular compounds may be specified at different levels of abstraction. At the coarsest, arithmetical level molecular formulas describe only the number and type of constituent atoms; a finer topological level uses graphs to determine the adjacencies between atoms; a further refinement also determines the (relative) spatial arrangement of atoms and thus the molecule's geometry. Stereoisomers, that is, molecules with the same topology but different geometry, often have similar physical and chemical properties but differ dramatically in their biological and pharmacological activity. A famous example is the sedative thalidomide. The compound with the German trade name Contergan has a sedative effect. Its non-superposable mirror image (such a pair of compounds is called enantiomorphic), however, causes severe birth defects. The stereospecific — and in particular enantioselective — synthesis of such compounds is a very challenging task in practice. In order for a graph transformation model of chemistry to be useful in practical applications, it therefore needs to be able to properly model stereoisomers and stereospecific chemical reactions. This task is not made simpler by the fact that stereochemical terms are often not well-defined in a mathematical sense [16].

To date, most chemical graph transformation models, with the notable exception is the hypergraph rewriting approach explored in [10], lack support for stereochemistry. The chemical literature, however, has recognized early-on that the ability to handle stereochemistry is a prerequisite for the practical applicability of computational models of chemistry: Already in the sixties of the last century the "ordered list method" was introduced [20,28]. It exploits the representation of graphs as adjacency lists by using the ordering of the edge lists to encode geometric information. Alternative approaches rely on transformation of structures to larger, ordinary graphs that encode the stereochemical situations, e.g. [1] or aim at the encoding in the form of linear descriptions such as SMILES [27] or CAST [25]. The chemical literature usually annotates local geometric information in terms of IUPAC nomenclature rules. For example, the local geometry at a tetrahedral centre is determined as "R" or "S" depending on a complex set of inherently non-local precedence rules for the four neighbours [8]. Such representation of geometric information is not designed to allow the implementation of chemical reactions as local rewriting operations.

Here we advocate a strategy that differs in a conceptually important point from [10]: Their hypergraph approach explicitly uses transformation rules to generate equivalent tetrahedral centres, which results in exponentially many graphs (in the number of centres) representing the same molecule. Instead, we propose here to incorporate the symmetries that define equivalent local geometries directly into the morphisms themselves. This also allows us to preserve the modelling principle that each graph is equivalent to just one molecule, and that each direct direction is a proper chemical reaction. It is not in all reactions that the (full) geometric information is relevant, and we therefore also introduce a hierarchy of local atom configurations that allows the representation of partially known stereo-information, both in graphs and rules. This approach can be seen as a special case of graph transformation with node inheritance [18], though we opt for a more direct modelling approach, closer to a practical implementation, where the inheritance is capture in an specialised algebra using principles from term algebras.

We introduce stereochemistry and molecular shapes in Sect. 2, and in Sect. 3 we describe the graph model and transformation system with attributes that encodes information about local geometry. We give several application examples in Sect. 4 and conclude with Sect. 5. In the Appendix we present the code used for the application examples.

2   Molecular Shapes

The connectivity of molecules can be modelled trivially by undirected graphs, but this ignores the relative placement of atoms and their neighbours in 3D space. An intermediary view is to look locally at each atom and characterise the shape that the incident bonds form. Each atom features (depending on its type) a certain number of valence electrons. Part of these are shared with adjacent atoms in the formation of chemical bonds, while others remain localized at their atom and form so-called lone electron pairs. The Lewis diagram [19] of a molecule describes the distribution of valence electrons into bonding electron pairs and lone pairs. Backed by a grounding in quantum theory, the Gillespie-Nyholm theory, also called the Valence-Shell Electron-Pair Repulsion (VSEPR) theory [15], then explains the local geometry in terms of Lewis formula by means of three simple rules: (1) electron pairs repulse each other and thus attain a geometry that maximizes their mutual angular distances; (2) double and triple bonds can be treated like single bonds; and (3) lone electron pairs are treated like chemical bonds. Changes in bond orders and/or the number of lone pairs therefore affect the geometry as part of a chemical reaction. The distinction between bonds and lone pair allows the model to define fine-grained shapes, for example:

- The oxygen in a water molecule has 2 lone pairs and 2 incident bonds, giving it the "bent" shape.
- The nitrogen in an ammonia molecule has 1 lone pair and 3 incident bonds, giving it the "trigonal pyramidal" shape.
- The carbon in a methane molecule has no lone pairs and 4 incident bonds, giving the "tetrahedral" shape.

In terms of the VSEPR theory, each of these three examples correspond to a central atom with four neighbours, and the difference in shape arise from distinguishing bonds from lone pairs. Two atoms with the same sum of incident bonds and lone pairs have the same intrinsic geometry, in this case as a tetrahedron with the atom in the centre and the neighbours placed in the corners. In the model we thus only consider the basic shapes, from which the "visible" geometry of the molecule can be recovered by considering the lone pairs.

A comprehensive model of stereochemistry should include separate treatment of each possible shape. In this contribution we focus on the tetrahedral shape and the general modelling framework that also allows for partial specification of stereo-information in transformation rules. Future extensions will then implement the remaining chemically relevant shapes.

Throughout the paper we use the depiction of tetrahedral shapes usually used in chemistry, where wedge (◀━━━) and hash (⋯⋯⫼⫼⫼) bonds are used to indicate their 3D embedding. In Fig. 1 this is illustrated on the two stereoisomers of glyceraldehyde.

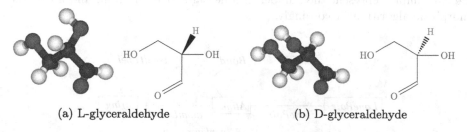

(a) L-glyceraldehyde   (b) D-glyceraldehyde

Fig. 1. Depiction of the two stereoisomers of glyceraldehyde in 3D (3D depictions from https://en.wikibooks.org/wiki/Organic_Chemistry/Chirality) and in 2D with wedge/hash bond notation to indicate the 3D embedding. The broad end of a wedge (resp. hash) bond is placed above (resp. below) the plane of drawing of the narrow end.

3   Model

3.1   Molecules as Typed Attributed Graphs

Molecules without stereochemical information can be modelled directly using simple undirected graphs, with labels on vertices and edges. For extending this model we recast the model described in [5] in terms of typed attributed graphs (e.g., see [9]), which simply results in the type graph shown in Fig. 2. In the practical use of a chemical graph transformation system it is useful to enable/disable stereochemical information in different contexts. The stereochemical model therefore only adds to the type graph of the basic model.

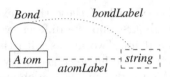

Fig. 2. Type graph for the basic molecule model, where each atom vertex and bond edge are attributed with strings, that encode the atom type, charge, and bond order.

Not all combinations of atom types, charges, number of lone pairs, and shapes are chemically valid. However, for simplicity we here present a general model for describing local geometry, and leave out the details of checking for chemical validity. The number of combinations is quite limited and in the end the check can therefore be handled by a moderately sized lookup table.

For representing lone pairs we allow each atom to have additional neighbours of type *LonePair* (see Fig. 3). In the following when we refer to the degree of an atom and its neighbours we thus include the lone pairs. On a practical note, we can simply represent the number of lone pairs at each atom, and adapt the morphism algorithms accordingly.

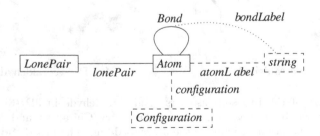

Fig. 3. The extended type graph for representing stereochemistry. A new type of vertex is introduced for the modelling of lone electron pairs, and a new atom attribute is added for representing molecular shapes and embeddings into the shapes. Each atom is only allowed to have 1 configuration, while it may have multiple neighbouring lone pairs.

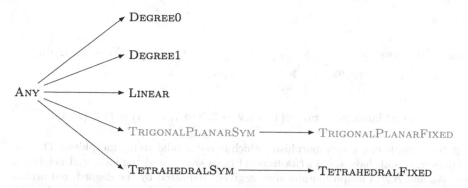

Fig. 4. The category of shapes, $\mathcal{C}_{\text{Shape}}$, used as a basis for encoding stereochemical configurations. Leaf objects correspond to actual molecular shapes while the remaining objects provide a means for specifying partial stereo-information by acting as "variable" shapes. In particular, the ANY shape is the initial object that acts as an unconstrained variable. The two trigonal planar shapes are shown only as an example of how the category will be extended in the future. They are briefly discussed in the concluding remarks.

Next we introduce a category of shapes $\mathcal{C}_{\text{Shape}}$, where the objects and morphisms are explicitly defined, see Fig. 4. In principle we add an object for each general shape described in the VSEPR theory, though here we focus on the tetrahedral shape. We additionally introduce several "variable" shapes for more expressive modelling of transformation rules, including the ANY shape which is the initial object of the category. This allows for the direct expression of (partially) unknown configurations, both in rules and in molecules.

In contrast to the "ordered list method" in [20, 28] we do not modify the underlying storage of the graph. Instead we store the neighbour ordering in a *Configuration* attribute on each atom along with the geometric shape of the atom. That is, a configuration is a pair $\langle S, N \rangle$ of a shape object S and an ordered list of all neighbours of the atom N. Most shapes may only be assigned to atoms of a specific degree (see below), e.g., the tetrahedral shape requires the atom to have degree 4. As each configuration references the neighbours in the graph, the definition of configuration morphisms requires an already valid graph morphism, which we assume also to be injective due to the modelling of chemistry [5]. Let $m \colon G_1 \to G_2$ be such an injective typed graph morphism, with respect to all attributes except for the configurations. For deciding whether m is also valid when taking configurations into account, consider an atom vertex u of G_1 with configuration $\langle S_1, N_1 \rangle$, and its image $v = m(u)$ with configuration $\langle S_2, N_2 \rangle$. We first require that a shape morphism $S_1 \to S_2$ exists. Then, from the neighbour lists $N_1 = [u_1, u_2, \ldots, u_{d_1}]$ and $N_2 = [v_1, v_2, \ldots, v_{d_2}]$ create an index map $m_I \colon \{1, 2, \ldots, d_1\} \to \{1, 2, \ldots, d_2\}$ such that if $m(u_i) = v_j$ then $m_I(i) = j$. Each shape morphism $S_1 \to S_2$ may now define additional constraints the index map m_I must fulfil (see Fig. 5 for an example). Though, for the current set of shapes only morphisms among configurations with TETRAHEDRALFIXED shape has additional constraints.

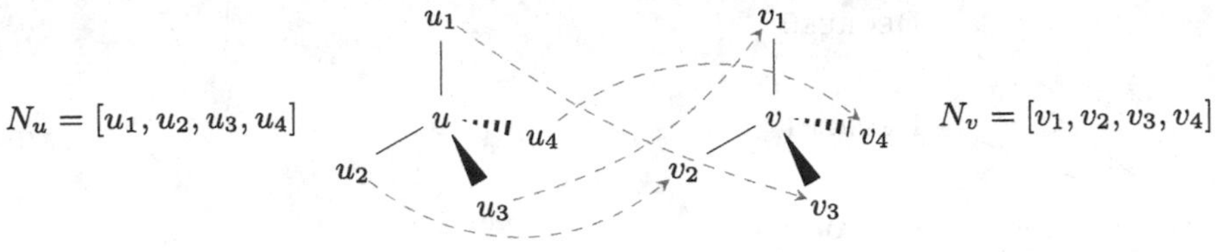

$$N_u = [u_1, u_2, u_3, u_4] \qquad\qquad\qquad\qquad\qquad N_v = [v_1, v_2, v_3, v_4]$$

Induced index map: $m_I = \{1 \mapsto 3, 2 \mapsto 2, 3 \mapsto 1, 4 \mapsto 4\} = (1\ 3)(2)(4)$

Fig. 5. Example of a graph morphism, which is not a valid stereo morphism. The two vertices u, v both have the TETRAHEDRALFIXED shape, and the indicated neighbour lists N_u and N_v. A graph morphism m is given, indicated by the dashed, red arrows and with $m(u) = v$. This induces the index map $m_I = \{1 \mapsto 3, 2 \mapsto 2, 3 \mapsto 1, 4 \mapsto 4\}$, i.e., the permutation $(1\ 3)(2)(4)$. As this permutation does not describe a symmetry of a tetrahedron, following our encoding convention, the graph morphism is not a valid stereo morphism. (Color figure online)

In the following we describe intended semantics, degree constraints, and index map constraints of each shape.

The TETRAHEDRALFIXED *Shape* can only be attached to atoms of degree 4. We interpret a neighbour list $[v_1, v_2, v_3, v_4]$ geometrically in the following manner: the neighbours are placed in the corners of a regular tetrahedron, and v is placed in the centre. When looking from v_1 towards v, the neighbours v_2, v_3, v_4 appear in counter-clockwise order. With this encoding the symmetries of a tetrahedron can be expressed as the permutation group generated by $\langle (1)(2\ 3\ 4), (1\ 2)(3\ 4)\rangle$ acting on the neighbour list, corresponding to the alternating group on 4 elements as expected. A morphism from one TETRAHEDRALFIXED configuration to another thus requires the index map to be a permutation from this group. In Fig. 5 an example of a graph morphism that does not meet this requirement is shown.

The TETRAHEDRALSYM *Shape.* In some cases the specific embedding of an atom in tetrahedral shape is unknown, and in some cases it is beneficial to be able to match both possible tetrahedral embeddings. We therefore introduce this shape that also requires atom degree 4, has the geometric shape of a tetrahedron, but with no particular assignment of neighbours to the corners. The symmetries of the neighbours are therefore the complete symmetric group on 4 elements. As it has a morphism to the TETRAHEDRALFIXED shape it can be used as a restricted "variable" in transformation rules.

The ANY *Shape* has no degree constraints, and all neighbour lists are equivalent. It is the initial object of the shape category, and can therefore be used as an unrestricted "variable" in transformation rules.

The DEGREE0, DEGREE1, *and* LINEAR *Shapes* require degree 0, 1, and 2, resp., of the atoms they are attached to. Geometrically, an atom with the LINEAR shape is located on the line between its two neighbours.

3.2 Transformation Rules and Derivations

For a DPO transformation rule $p = (L \xleftarrow{l} K \xrightarrow{r} R)$ we already require l and r to be graph monomorphisms. In the extension to stereochemical information, we require them to be isomorphisms on the configuration attributes. That is, either an atom has no configuration attribute in K, or it has the same attribute in L, K, and R. The top span of Fig. 6 shows an example rule where the change of configuration is combined with partial stereo-information. As configurations contain lists of neighbours in the graph, the isomorphism requirement for configurations implies that only atoms of K where all incident edges also are in K can have a configuration attribute. From the perspective of modelling chemistry this means that when bonds are broken or formed, one must be explicit about the change of molecular shape for the incident atoms.

In rule application the configurations with non-leaf shapes (see Fig. 4) act as unnamed variables, similar to transformation with term attributes described in [9]. That is, in the transformation of a graph G with a rule $p = (L \xleftarrow{l} K \xrightarrow{r} R)$, the match morphism $m\colon L \to G$ implicitly determines an assignment of configurations such that substitution yields isomorphic configurations. This is illustrated with both vertex 0 and 1 in the direct derivation shown in Fig. 6. Vertex 1 has an ANY configuration in L, and is being assigned to a TETRAHEDRALSYM configuration through m. As it also has this configuration in K and R, the pushout requirements preserve the TETRAHEDRALSYM configuration through D to H. Vertex 0 has a TETRAHEDRALSYM configuration in L, which is being assigned to a TETRAHEDRALFIXED configuration. However, the vertex has no configuration in K, and a new TETRAHEDRALSYM configuration is added in R. The rule therefore effectively matches any tetrahedron to vertex 0 and generalizes it to a TETRAHEDRALSYM.

4 Application Examples

We have extended the graph transformation system of MØD [2,5] with the model for stereochemistry. Morphisms are found using the VF2 algorithm [7], where shape morphisms are checked during matching. Index map constraints require the complete neighbourhood of a vertex to be mapped to the host graph. For simplicity this check is deferred to after a total morphism has been found.

In the following we illustrate the use of the modelling framework. The code for each example can be found in the appendix, and can be experimented with in the live version of MØD at http://mod.imada.sdu.dk/playground.html.

4.1 Stereospecific Aconitase

One of the central metabolic pathways is the citric acid cycle, which contains a reaction that converts the molecule citrate into isocitrate. This reaction, facilitated by the aconitase enzyme, is stereospecific which means that it only produces D-isocitrate and not the stereoisomer L-isocitrate. While the modelling of

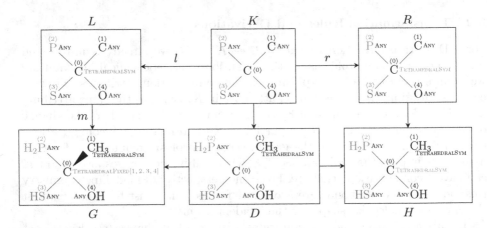

Fig. 6. A direct derivation with explicitly annotated configuration data. Vertex 0 and 1 have variable configurations with TETRAHEDRALSYM and ANY shape, such that they can match more specialised configurations. As vertex 1 also has a configuration in K, its assigned TETRAHEDRALSYM configuration in G is transferred to D and H as well. The configuration on vertex 0 is on the other hand being deleted and replaced with a new configuration in R. The original TETRAHEDRALFIXED configuration in G is therefore replaced accordingly.

this reaction as a transformation rule can be done in the hypergraph approach described in [10], the present approach also allows us to generalize the rule to be applicable to molecules other than isocitrate, that share the same context. This is shown in Fig. 7 where a generalized rule for aconitase is shown being applied to citrate and water.

4.2   Generation of Stereoisomers

Tartaric acid is the most important chemical compound for the discovery of the concept of chirality. Tartaric acid has three stereoisomers, two are chiral (i.e., their mirror image is non-superposable) and one is achiral (i.e., it equals its mirror image). The crystal structure of the double salt of the stereoisomers of tartaric acid (potassium sodium tartrate tetrahydrate) was analysed by Louis Pasteur. He performed a morphological analysis and analysed the shapes of the different macroscopic crystals. The macroscopic (non-)superposability of the idealised shape of the crystals established the existence of molecular chirality [12].

We use the tartaric acid molecule here as an example to illustrate how all stereoisomers with partial and fully specified stereoinformation can be inferred in the rule-based framework. This is accomplished by repeated application of the rule shown in Fig. 8. As the central atom has TETRAHEDRALSYM shape it can be used to either fixate the tetrahedral embedding or change an existing one. We here also extend the ordinary atom labels to include the special unnamed variable label '*' that can be assigned any other atom label during matching. Figure 9 shows the result of repeatedly applying the rule to a model of tartaric

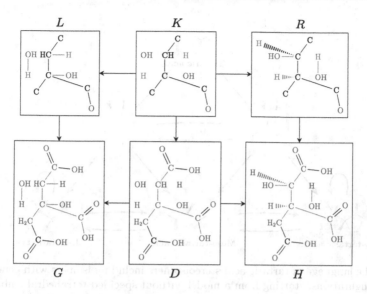

Fig. 7. Illustration of a generalized transformation rule for the aconitase enzyme, used in the citric acid cycle, applied to a citrate and water molecule. The reaction is stereospecific, and results therefore in D-isocitrate but not L-isocitrate. In the left side the two central carbon atoms have the TETRAHEDRALSYM shape, in order to match any tetrahedral, while in the right side they both have the more specialized TETRAHEDRAL-FIXED shape with a specific embedding.

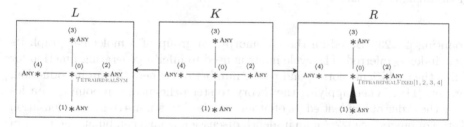

Fig. 8. A generic rule that either fixates or changes the embedding of a tetrahedral atom. Each vertex is annotated explicitly with the configuration data, and the asterisks * are unnamed variable labels that match any atom label.

acid without fully specified stereo-information. We see that the 3 stereoisomers, L- and meso-tataric acid, in addition to the naturally occurring form D-tartaric acid are generated as expected.

While it is not too difficult to manually derive the stereoisomers of tartaric acid, the task quickly becomes complicated and error prone for larger molecules. Enumeration of (i.e., explicitly creating all) and counting molecules has been providing a fertile ground for developments in graph theory, combinatorics, chemistry and the intersecting research fields since the nineteenth century. Many counting problems in chemistry have been solved by the Pólya Theory of

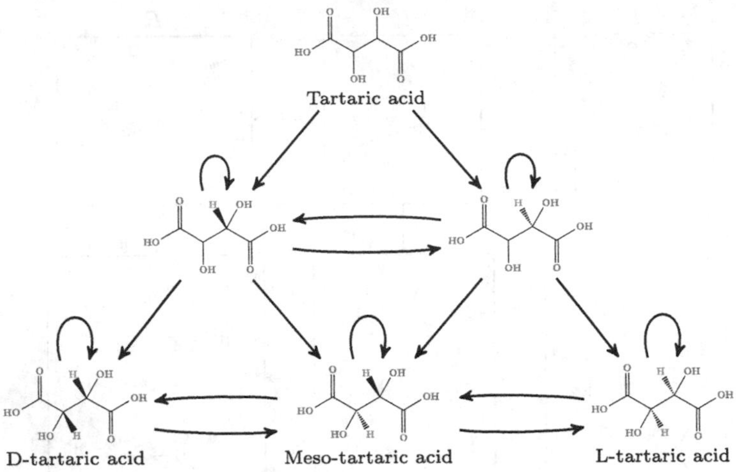

Fig. 9. The language of tartaric acid stereoisomers including isomers with generalized stereo configurations, starting from a model without specified tetrahedral embeddings (the graph on top). Each arrow represents a direct derivation using the rule shown in Fig. 8. As it matches any tetrahedral configuration, it also results in identity derivations for molecules already with a TETRAHEDRALFIXED atom. The bottom three graphs, D-, L-, and meso-tartaric acid, are models with fully specified embeddings, and are therefore the proper stereoisomers. The two graphs in the middle only have a tetrahedral embedding fixated on one of the two central carbon atoms, while the other still has TETRAHEDRALSYM shape.

Counting [21,23]. Based on the automorphism group of a molecular graph its cycle index is inferred. The cycle index is used to infer a generating function for which the coefficients correspond to the number of isomers (for an introduction, e.g., see [11]). When applying the theory to stereochemical compounds, considering the order of incident edges of atoms can lead to a non-trivial compensation of stereoisomers (see [22] for an in-depth discussion from a combinatorial point of view). An example is shown in Fig. 10, where a central tetrahedral carbon atom (adjacent to the nitrogen atom) has two graph-isomorphic subtrees attached, i.e., they are isomorphic if the stereo information is ignored. If two different tetrahedral embeddings are added to the subtree carbons, then the central carbon atom can have only one tetrahedral embedding up to isomorphism (the outer graphs of Fig. 10). On the other hand, when two different embeddings are on the subtree carbons, then only two further stereoisomers exist (the inner graphs); one for each of the embeddings on the central carbon. This kind of compensation of such stereoisomers has been thoroughly analysed for specific molecular classes (e.g. tree-like structures with single bonds only) in literature. However, our framework allows not only for enumeration of stereoisomers, but also for a rigorous modelling of chemical and biochemical pathways with complete or partial stereoinformation attached.

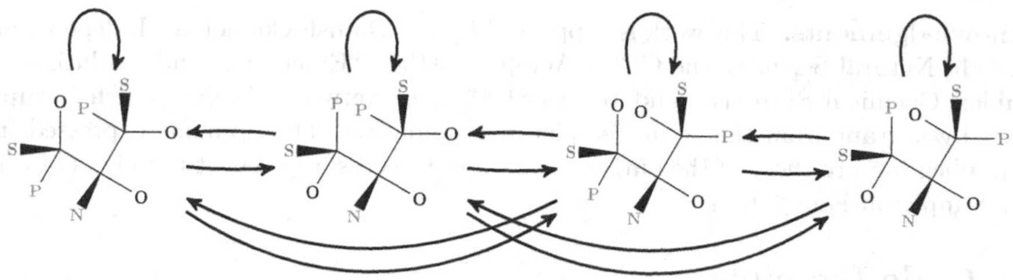

Fig. 10. The language of all proper stereoisomers for an abbreviated molecule, using the rule shown in Fig. 8. As with the tartaric acid example (Fig. 9) the rule can result in identity derivations. All three carbon atoms have TETRAHEDRALFIXED shape.

5 Concluding Remarks

We have presented a model of molecules based on typed attributed graphs that include the representation of local molecular shapes. The model is inspired by previous work on molecule representation, e.g., the ordered list method from chemistry and the hypergraph approach from graph transformation. We have extended it here to allow a partial specification of stereochemical information. This both allows for partially assigning geometric information to molecules, but more importantly provides a more expressive framework for describing classes of reactions as graph transformation rules. The presented model additionally includes the possibility to represent lone electron pairs, which in some cases give rise to multiple stereoisomers. We have implemented the model as an extension of the chemical graph transformation system in the MØD software package. The extension is being prepared for release in an upcoming version of MØD.

Additional Shapes. The trigonal planar shape is another important shape in biochemistry, which gives rise to cis-trans isomerism in conjunction with incident double or aromatic bonds. In this shape an atom is coplanar with its required 3 neighbours. In Fig. 4 we have shown how this shape can be added to the shape category. Like the TETRAHEDRALFIXED shape, it has associated constraints on index maps induced by graph morphisms. In addition the trigonal planar shape will also require non-local checks of morphisms to ensure consistency of the half-planes implicitly defined by the neighbour lists.

Shapes that require more than 4 neighbours are uncommon in biochemistry, although the trigonal bipyramid plays a role in phosphorus chemistry. Preliminary investigations suggest that all other chemically relevant local geometries can also be defined in the framework laid out in this contribution.

The embedding of a graph in the plane (or any surface) can be represented by locally imposing a cyclic order on the incident edges at each vertex, also called a rotation system. The semantics of this encoding is similar to that of the trigonal planar shape. The same techniques thus are applicable to defining a transformation system for graphs with an associated embedding.

Acknowledgements. This work is supported by the Danish Council for Independent Research, Natural Sciences, the COST Action CM1304 "Emergence and Evolution of Complex Chemical Systems", and the ELSI Origins Network (EON), which is supported by a grant from the John Templeton Foundation. The opinions expressed in this publication are those of the authors and do not necessarily reflect the views of the John Templeton Foundation.

A   Code Examples

The following code shows how to use the stereochemical extension of MØD, in the context of the three application examples. The code is also available as modifiable scripts in the live version of the software, accessible at http://mod. imada.sdu.dk/playground.html.

A.1   Stereospecific Aconitase

Executing the following code creates the figures for Fig. 7.

```
water = smiles("O", "H_2O")
cit = smiles("C(C(C(=O)O)C(CC(=O)O)(C(=O)O)O", name="Cit")
d_icit = smiles("C([C@@H]([C@H](C(=O)O)O)C(=O)O)C(=O)O", name="D-ICit")

aconitase = ruleGMLString("""rule [
    left [
        # the dehydrated water
        edge [ source 1 target 100 label "-"] edge [ source 2 target 102 label "-"]
        # the hydrated water
        edge [ source 200 target 202 label "-"]
    ]
    context [
        node [ id 1 label "C"]
        edge [ source 1 target 2 label "-"] # goes from - to = to -
        node [ id 2 label "C"]
        # the dehydrated water
        node [ id 100 label "O"] node [ id 101 label "H"] node [ id 102 label "H"]
        edge [ source 100 target 101 label "-"]
        # the hydrated water
        node [ id 200 label "O"] node [ id 201 label "H"] node [ id 202 label "H"]
        edge [ source 200 target 201 label "-"]
        # dehydrated C neighbours
        node [ id 1000 label "C"] node [ id 1010 label "O"] node [ id 1001 label "C"]
        edge [ source 1 target 1000 label "-"] edge [ source 1000 target 1010 label "-"]
        edge [ source 1 target 1001 label "-"]
        # hydrated C neighbours
        node [ id 2000 label "C"] node [ id 2001 label "H"]
        edge [ source 2 target 2000 label "-"] edge [ source 2 target 2001 label "-"]
    ]
    right [
        # The '!' in the end changes it from TetrahedralSym to
        # TetrahedralFixed
        node [ id 1 stereo"tetrahedral[1000, 1001, 202, 2]!"]
        node [ id 2 stereo"tetrahedral[200, 1, 2000, 2001]!"]
        # the dehydrated water
        edge [ source 100 target 102 label "-"]
        # the hydrated water
        edge [ source 1 target 202 label "-"] edge [ source 2 target 200 label "-"]
    ]
]""")

dg = dgRuleComp(inputGraphs, addSubset(cit, water) >> aconitase,
        # seldctino of attributes and morphisms for matching
        labelSettings=LabelSettings(
            # use terms as labels, instead of strings
            LabelType.Term,
            # term morphisms may be specialisations
            LabelRelation.Specialisation,
            # use stereo information,
            # with specialisation in the morphisms
            LabelRelation.Specialisation)
)
dg.calc()
for e in dg.edges:
    p = GraphPrinter()
    p.withColour = True
    e.print(p, matchColour="Maroon")
```

A.2   Stereoisomers of Tartaric Acid

Executing the following code creates the figures for Figs. 8 and 9.

```
smiles("C(C(C(=0)0)0)(C(=0)0)0", name="Tartaric acid")
smiles("[C@@H]([C@H](C(=0)0)0)(C(=0)0)0", name="L-tartaric acid")
smiles("[C@H]([C@@H](C(=0)0)0)(C(=0)0)0", name="D-tartaric acid")
smiles("[C@@H]([C@@H](C(=0)0)0)(C(=0)0)0", name="Meso-tartaric acid")
change = ruleGMLString("""rule [
    left    [   node [ id 0 stereo"tetrahedral"] ]
    context [
        node [ id 0 label "*"] node [ id 1 label "*"] node [ id 2 label "*"]
        node [ id 3 label "*"] node [ id 4 label "*"]
        edge [ source 0 target 1 label "-"]
        edge [ source 0 target 3 label "-"] edge [ source 0 target 4 label "-"]
    ]
    right   [   node [ id 0 stereo"tetrahedral[1, 2, 3, 4]!"] ]
]""")

dg = dgRuleComp(inputGraphs, addSubset(inputGraphs) >> repeat(change),
    # seldctino of attributes and morphisms for matching
    labelSettings=LabelSettings(
        # use terms as labels, instead of strings
        LabelType.Term,
        # term morphisms may be specialisations
        LabelRelation.Specialisation,
        # use stereo information,
        # with specialisation in the morphisms
        LabelRelation.Specialisation)
)
dg.calc()

p = GraphPrinter()
p.setMolDefault()
p.withPrettyStereo = True
change.print(p)
p = DGPrinter()
p.withRuleName = True
p.withRuleId = False
dg.print(p)
```

A.3   Non-trivial Stereoisomers

Executing the following code creates the figures for Figs. 8 and 10.

```
g = smiles("[N][C@]([O])([C@]([S])([P])([O])([C@]([S])([P])([O])))")
change = ruleGMLString("""rule [
    left    [   node [ id 0 stereo"tetrahedral"] ]
    context [
        node [ id 0 label "*"] node [ id 1 label "*"] node [ id 2 label "*"]
        node [ id 3 label "*"] node [ id 4 label "*"]
        edge [ source 0 target 1 label "-"] edge [ source 0 target 2 label "-"]
        edge [ source 0 target 3 label "-"] edge [ source 0 target 4 label "-"]
    ]
    right   [   node [ id 0 stereo"tetrahedral[1, 2, 3, 4]!"] ]
]""")

dg = dgRuleComp(inputGraphs, addSubset(inputGraphs) >> repeat(change),
    # seldctino of attributes and morphisms for matching
    labelSettings=LabelSettings(
        # use terms as labels, instead of strings
        LabelType.Term,
        # term morphisms may be specialisations
        LabelRelation.Specialisation,
        # use stereo information,
        # with specialisation in the morphisms
        LabelRelation.Specialisation)
)
dg.calc()

p = GraphPrinter()
p.setMolDefault()
p.withPrettyStereo = True
change.print(p)
p = DGPrinter()
p.withRuleName = True
p.withRuleId = False
dg.print(p)
```

References

1. Akutsu, T.: A new method of computer representation of stereochemistry. Transforming a stereochemical structure into a graph. J. Chem. Inf. Comput. Sci. **31**, 414–417 (1991)
2. Andersen, J.L.: MedØlDatschgerl (MØD) (2016). http://mod.imada.sdu.dk
3. Andersen, J.L., Flamm, C., Merkle, D., Stadler, P.F.: Inferring chemical reaction patterns using rule composition in graph grammars. J. Syst. Chem. **4**(1), 4 (2013)
4. Andersen, J.L., Flamm, C., Merkle, D., Stadler, P.F.: Generic strategies for chemical space exploration. Int. J. Comput. Biol. Drug Des. **7**(2/3), 225–258 (2014). http://arxiv.org/abs/1302.4006
5. Andersen, J.L., Flamm, C., Merkle, D., Stadler, P.F.: A software package for chemically inspired graph transformation. In: Echahed, R., Minas, M. (eds.) ICGT 2016. LNCS, vol. 9761, pp. 73–88. Springer, Cham (2016). doi:10.1007/978-3-319-40530-8_5
6. Benkö, G., Flamm, C., Stadler, P.F.: A graph-based toy model of chemistry. J. Chem. Inf. Comput. Sci. **43**, 1085–1093 (2003)
7. Cordella, L., Foggia, P., Sansone, C., Vento, M.: A (sub) graph isomorphism algorithm for matching large graphs. IEEE Trans. Pattern Anal. Mach. Intell. **26**(10), 1367 (2004)
8. Cross, L.C., Klyne, W.: Rules for the nomenclature of organic chemistry: section E: stereochemistry. Pure Appl. Chem. **45**, 11–30 (1976)
9. Ehrig, H., Ehrig, K., Prange, U., Taenthzer, G.: Fundamentals of Algebraic Graph Transformation. Springer, Berlin (2006)
10. Ehrig, K., Heckel, R., Lajios, G.: Molecular analysis of metabolic pathway with graph transformation. In: Corradini, A., Ehrig, H., Montanari, U., Ribeiro, L., Rozenberg, G. (eds.) ICGT 2006. LNCS, vol. 4178, pp. 107–121. Springer, Heidelberg (2006). doi:10.1007/11841883_9
11. Faulon, J.L., Visco Jr., D., Roe, D.: Enumerating Molecules, Reviews in Computational Chemistry, vol. 21, pp. 209–286. Wiley, Hoboken (2005)
12. Flack, H.D.: Louis Pasteur's discovery of molecular chirality and spontaneous resolution in 1848, together with a complete review of his crystallographic and chemical work. Acta Crystallogr. Sect. A **65**, 371–389 (2009)
13. Fontana, W., Buss, L.W.: "The arrival of the fittest": toward a theory of biological organization. Bull. Math. Biol. **56**, 1–64 (1994)
14. Fontana, W., Buss, L.W.: What would be conserved "if the tape were played twice". Proc. Natl. Acad. Sci. USA **91**, 757–761 (1994)
15. Gillespie, R.: Fifty years of the VSEPR model. Coord. Chem. Rev. **252**, 1315–1327 (2008)
16. Kerber, A., Laue, R., Meringer, M., Rücker, C., Schymanski, E.: Mathematical Chemistry and Chemoinformatics. De Gruyter (2013)
17. Kreowski, H.J., Kuske, S.: Graph multiset transformation: a new framework for massively parallel computation inspired by DNA computing. Nat. Comput. **10**(2), 961–986 (2011)
18. de Lara, J., Bardohl, R., Ehrig, H., Ehrig, K., Prange, U., Taentzer, G.: Attributed graph transformation with node type inheritance. Theor. Comput. Sci. **376**(3), 139–163 (2007). http://www.sciencedirect.com/science/article/pii/S0304397507000631
19. Lewis, G.N.: The atom and the molecule. J. Am. Chem. Soc. **38**, 762–785 (1916)

20. Petrarca, A.E., Lynch, M.F., Rush, J.E.: A method for generating unique computer structural representation of stereoisomers. J. Chem. Doc. **7**, 154–165 (1967)
21. Pólya, G.: Kombinatorische Anzahlbestimmungen für Gruppen, Graphen und chemische Verbindungen. Acta Mathematica **68**(1), 145–254 (1937)
22. Pólya, G., Read, R.: Combinatorial Enumeration of Groups, Graphs, and Chemical Compounds. Springer, New York (1987)
23. Redfield, J.: The theory of group-reduced distributions. Am. J. Math. **49**(3), 433–455 (1927)
24. Rosselló, F., Valiente, G.: Graph transformation in molecular biology. In: Kreowski, H.-J., Montanari, U., Orejas, F., Rozenberg, G., Taentzer, G. (eds.) Formal Methods in Software and Systems Modeling. LNCS, vol. 3393, pp. 116–133. Springer, Heidelberg (2005). doi:10.1007/978-3-540-31847-7_7
25. Satoh, H., Koshino, H., Funatsu, K., Nakata, T.: Novel canonical coding method for representation of three-dimensional structures. J. Chem. Inf. Comput. Sci. **40**, 622–630 (2000)
26. Taentzer, G.: AGG: a graph transformation environment for modeling and validation of software. In: Pfaltz, J.L., Nagl, M., Böhlen, B. (eds.) Applications of Graph Transformations with Industrial Relevance: Second International Workshop, AGTIVE 2003. LNCS, vol. 3062, pp. 446–453. Springer, Heidelberg (2004)
27. Weininger, D.: SMILES, a chemical language and information system. 1. Introduction to methodology and encoding rules. J. Chem. Inf. Model. **28**, 31–36 (1988)
28. Wipke, W.T., Dyott, T.M.: Simulation and evaluation of chemical synthesis-computer representation and manipulation of stereochemistry. J. Am. Chem. Soc. **96**, 4825–4834 (1974)
29. Yadav, M.K., Kelley, B.P., Silverman, S.M.: The potential of a chemical graph transformation system. In: Ehrig, H., Engels, G., Parisi-Presicce, F., Rozenberg, G. (eds.) ICGT 2004. LNCS, vol. 3256, pp. 83–95. Springer, Heidelberg (2004). doi:10.1007/978-3-540-30203-2_8

Graph Languages and Parsing

Specifying Graph Languages with Type Graphs

Andrea Corradini[1], Barbara König[2], and Dennis Nolte[2(✉)]

[1] Università di Pisa, Pisa, Italy
andrea@di.unipi.it
[2] Universität Duisburg-Essen, Duisburg, Germany
{barbara_koenig,dennis.nolte}@uni-due.de

Abstract. We investigate three formalisms to specify graph languages, i.e. sets of graphs, based on type graphs. First, we are interested in (pure) type graphs, where the corresponding language consists of all graphs that can be mapped homomorphically to a given type graph. In this context, we also study languages specified by restriction graphs and their relation to type graphs. Second, we extend this basic approach to a type graph logic and, third, to type graphs with annotations. We present decidability results and closure properties for each of the formalisms.

1 Introduction

Formal languages in general and regular languages in particular play an important role in computer science. They can be used for pattern matching, parsing, verification and in many other domains. For instance, verification approaches such as reachability checking, counterexample-guided abstraction refinement [5] and non-termination analysis [11] could be directly adapted to graph transformation systems if one had a graph specification formalism with suitable closure properties, computable pre- and postconditions and inclusion checks. Inclusion checks are also important for checking when a fixpoint iteration sequence stabilizes.

While regular languages for words and trees are well-understood and can be used efficiently and successfully in applications, the situation is less satisfactory when it comes to graphs. Although the work of Courcelle [9] presents an accepted notion of recognizable graph languages, equivalent to regular languages, this is often not useful in practice, due to the sheer size of the resulting graph automata. Other formalisms, such as application conditions [13,20] and first-order or second-order logics, feature more compact descriptions, but there are problems with expressiveness, undecidability issues or unsatisfactory closure properties.[1]

Hence, we believe that it is important to study and compare specification formalisms (i.e., automata, grammars and logics) that allow to specify potentially infinite sets of graphs. In our opinion there is no one-fits-all solution, but we believe that specification mechanisms should be studied and compared more extensively.

---

[1] A more detailed overview over related formalisms is given in the conclusion (Sect. 6).

© Springer International Publishing AG 2017
J. de Lara and D. Plump (Eds.): ICGT 2017, LNCS 10373, pp. 73–89, 2017.
DOI: 10.1007/978-3-319-61470-0_5

In this paper we study specification formalisms based on type graphs, where a type graph T represents all graphs that can be mapped homomorphically to T, potentially taking into account some extra constraints. Type graphs are common in graph rewriting [7, 21]. Usually, one assumes that all items, i.e., rules and graphs to be rewritten, are typed, introducing constraints on the applicability of rules. Hence, type graphs are in a way seen as a form of labelling. This is different from our point of view, where graphs (and rules) are – a priori – untyped (but labeled) and type graphs are simply a means to represent sets of graphs.

There are various reasons for studying type graphs: first, they are reasonably simple with many positive decidability results and they have not yet been extensively studied from the perspective of specification formalisms. Second, other specification mechanisms – especially those used in connection with verification and abstract graph transformation [19, 23, 24] – are based on type graphs: abstract graphs are basically type graphs with extra annotations. Third, while not being as expressive as recognizable graph languages, they retain a nice intuition from regular languages: given a finite state automaton M one can think of the language of M as the set of all string graphs that can be mapped homomorphically to M (respecting initial and final states).

We in fact study three different formalisms based on type graphs: first, pure type graphs T, where the language consists simply of all graphs that can be mapped to T. We also discuss the connection between type graph and restriction graph languages. Then, in order to obtain a language with better boolean closure properties, we study type graph logic, which consists of type graphs enriched with boolean connectives (negation, conjunction, disjunction). Finally, we consider annotated type graphs, where the annotations constrain the number of items mapped to a specific node or edge, somewhat similar to the proposals from abstract graph rewriting mentioned above.

In all three cases we are interested in closure properties and in decidability issues (such decidability of the membership, emptiness and inclusion problems) and in expressiveness. Proofs for all the results and an extended example for annotated type graphs can be found in [6].

2 Preliminaries

We first introduce graphs and graph morphisms. In the context of this paper we use edge-labeled, directed graphs.

Definition 1 (Graph). *Let Λ be a fixed set of edge labels. A Λ-labeled graph is a tuple $G = \langle V, E, src, tgt, lab \rangle$, where V is a finite set of nodes, E is a finite set of edges, $src, tgt \colon E \to V$ assign to each edge a source and a target node, and $lab \colon E \to \Lambda$ is a labeling function.*

We will denote, for a given graph G, its components by V_G, E_G, src_G, tgt_G and lab_G, unless otherwise indicated.

Definition 2 (Graph morphism). *Let G, G' be two Λ-labeled graphs. A graph morphism $\varphi\colon G \to G'$ consists of two functions $\varphi_V\colon V_G \to V_{G'}$ and $\varphi_E\colon E_G \to E_{G'}$, such that for each edge $e \in E_G$ it holds that $src_{G'}(\varphi_E(e)) = \varphi_V(src_G(e))$, $tgt_{G'}(\varphi_E(e)) = \varphi_V(tgt_G(e))$ and $lab_{G'}(\varphi_E(e)) = lab_G(e)$. If φ is both injective and surjective it is called an isomorphism.*

We will often drop the subscripts V, E and write φ instead of φ_V, φ_E. We will consider the category **Graph** having Λ-labeled graphs as objects and graph morphisms as arrows. The set of its objects will be denoted by $\mathbf{Gr}_\Lambda$. The categorical structure induces an obvious preorder on graphs, defined as follows.

Definition 3 (Homomorphism preorder). *Given graphs G and H, we write $G \to H$ if there is a graph morphism from G to H in **Graph**. The relation $\to$ is obviously a preorder (i.e. it is reflexive and transitive) and we call it the homomorphism preorder on graphs. We write $G \nrightarrow H$ if $G \to H$ does not hold. Graphs G and H are homomorphically equivalent, written $G \sim H$, if both $G \to H$ and $H \to G$ hold.*

We will revisit the concept of *retracts* and *cores* from [15]. *Cores* are a convenient way to minimize type graphs, as, according to [15], all graphs G, H with $G \sim H$ have isomorphic cores.

Definition 4 (Retract and core). *A graph H is called a retract of a graph G if H is a subgraph of G and in addition there exists a morphism $\varphi\colon G \to H$. A graph H is called a core of G, written $H = core(G)$, if it is a retract of G and has itself no proper retracts.*

Example 5. The graph H is a retract of G, where the morphism φ is indicated by the node numbering:

Since the graph H does not have a proper retract itself it is also the core of G.

3   Languages Specified by Type or Restriction Graphs

In this section we introduce two classes of graph languages that are characterized by two somewhat dual properties. A *type graph language* contains all graphs that can be mapped homomorphically to a given *type graph*, while a *restriction graph language* includes all graphs that *do not contain* an homomorphic image of a given *restriction graph*. Next, we discuss for these two classes of languages some properties such as closure under set operators, decidability of emptiness and inclusion, and decidability of closure under rewriting via double-pushout rules. Finally we discuss the relationship between these two classes of graph languages.

Definition 6 (Type graph language). *A type graph T is just a Λ-labeled graph. The language $\mathcal{L}(T)$ is defined as:*

$$\mathcal{L}(T) = \{G \mid G \to T\}.$$

Example 7. The following type graph T over the edge label set $\Lambda = \{A, B\}$ specifies a type graph language $\mathcal{L}(T)$ consisting of infinitely many graphs:

The category **Graph** has a final object, that we denote $T_{\mathbf{*}}^{\Lambda}$, consisting of one node (called *flower node* $\mathbf{*}$) and one loop for each label in Λ. Therefore $\mathcal{L}(T_{\mathbf{*}}^{\Lambda}) = \mathbf{Gr}_{\Lambda}$. The graph $T_{\mathbf{*}}^{\Lambda}$ for $\Lambda = \{A, B, C\}$ is depicted to the right.

Specifying graph languages using type graphs gives us the possibility to forbid certain graph structures by not including them into the type graph. For example, no graph in the language of Example 7 can contain a B-loop or an A-edge incident to the target of a B-edge. However,

it is not possible to force some structures to exist in all graphs of the language, since the morphism to the type graph need not be surjective. This point will be addressed with the notion of *annotated type graph* in Sect. 5.

Another way (possibly more explicit) to specify languages of graphs not including certain structures, is the following one.

Definition 8 (Restriction graph language). *A restriction graph R is just a Λ-labeled graph. The language $\mathcal{L}_R(R)$ is defined as:*

$$\mathcal{L}_R(R) = \{G \mid R \nrightarrow G\}.$$

We will consider the relationship between the class of languages introduced in Definitions 6 and 8 in Sect. 3.3.

3.1 Closure and Decidability Properties

The type graph and restriction graph languages enjoy the following complementary closure properties with respect to set operators.

Proposition 9. *Type graph languages are closed under intersection (by taking the product of type graphs) but not under union or complementation, while restriction graph languages are closed under union (by taking the coproduct of restriction graphs) but not under intersection or complementation.*

Instead the two classes of languages enjoy similar decidability properties.

Proposition 10. *For a graph language $\mathcal{L}$ characterized by a type graph T (i.e. $\mathcal{L} = \mathcal{L}(T)$) or by a restriction graph R (i.e. $\mathcal{L} = \mathcal{L}_R(R)$) the following problems are decidable:*

1. *Membership, i.e. for each graph G it is decidable if $G \in \mathcal{L}$ holds.*
2. *Emptiness, i.e. it is decidable if $\mathcal{L} = \varnothing$ holds.*

Furthermore, language inclusion is decidable for both classes of languages:

3. *Given type graphs T_1 and T_2 it is decidable if $\mathcal{L}(T_1) \subseteq \mathcal{L}(T_2)$ holds.*
4. *Given restriction graphs R_1 and R_2 it is decidable if $\mathcal{L}_R(R_1) \subseteq \mathcal{L}_R(R_2)$ holds.*

3.2   Closure Under Double-Pushout Rewriting

In this subsection we are using the DPO approach with general, not necessarily injective, rules and matches. We discuss how we can show that a graph language $\mathcal{L}$ is a closed under a given graph transformation rule $\rho = (L \leftarrow \varphi_L - I - \varphi_R \rightarrow R)$, i.e., $\mathcal{L}$ is an invariant for ρ. This means that for all graphs G, H, where G can be rewritten to H via ρ, it holds that $G \in \mathcal{L}$ implies $H \in \mathcal{L}$.

For both type graph languages and restriction graph languages, separately, we characterize a sufficient and necessary condition which shows that closure under rule application is decidable. The condition for restriction graph languages is related to a condition already discussed in [14].

Proposition 11 (Closure under DPO rewriting for restriction graphs).
A restriction graph language $\mathcal{L}_R(S)$ is closed under a rule $\rho = (L \leftarrow \varphi_L - I - \varphi_R \rightarrow R)$ if and only if the following condition holds: for every pair of morphisms $\alpha\colon R \to F$, $\beta\colon S \to F$ which are jointly surjective, applying the rule ρ with (co-) match α backwards to F yields a graph E with a homomorphic image of S, i.e., $E \notin \mathcal{L}_R(S)$.

Proposition 12 (Closure under DPO rewriting for type graphs). *A type graph language $\mathcal{L}(T)$ is closed under a rule $\rho = (L \leftarrow \varphi_L - I - \varphi_R \rightarrow R)$ if and only if for each morphism $t_L\colon L \to core(T)$, there exists a morphism $t_R\colon R \to core(T)$ such that $t_L \circ \varphi_L = t_R \circ \varphi_R$.*

$$\mathcal{L}(T) \text{ is closed under application of } \rho \quad \Leftrightarrow \quad
\begin{array}{ccc}
 & \overset{\rho}{\overline{\phantom{L \xleftarrow{\varphi_L} I \xrightarrow{\varphi_R} R}}} & \\
L & \xleftarrow{\varphi_L} \; I \; \xrightarrow{\varphi_R} & R \\
\forall t_L \searrow & & \swarrow \exists t_R \\
 & core(T) &
\end{array}$$

We show that the *only if part* ($\Rightarrow$) of Proposition 12 cannot be weakened by considering morphisms to the type graph T, instead of to $core(T)$. In fact, consider the following type graph T and the rule ρ:

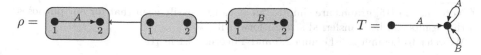

The type graph T contains the flower node, i.e., it has $T_{\ast}^{\{A,B\}}$ as subgraph. This ensures that each graph G, edge-labeled over $\Lambda = \{A, B\}$, is in the language $\mathcal{L}(T)$, and thus by rewriting any graph $G \in \mathcal{L}(T)$ into a graph H using ρ it is guaranteed that $H \in \mathcal{L}(T)$. However there is a morphism $t_L \colon L \to T$, the one mapping the A-labeled edge of L to the left A-labeled edge of T, such that there exists no morphism $t_R \colon R \to T$ satisfying $t_L \circ \varphi_L = t_R \circ \varphi_R$.

3.3   Relating Type Graph and Restriction Graph Languages

Both type graph and restriction graph languages specify collections of graphs by forbidding the presence of certain structures. This is more explicit with the use of restriction graphs, though. A natural question is how the two classes of languages are related. A partial answer to this is provided by the notion of *duality pairs* and by an important result concerning their existence, presented in [15].[2]

Definition 13 (Duality pair). *Given two graphs R and T, we call T the* dual *of R if for every graph G it holds that $G \to T$ if and only if $R \nrightarrow G$. In this case the pair (R, T) is called* duality pair.

Clearly, we have that (R, T) is a duality pair if and only if the restriction graph language $\mathcal{L}_R(R)$ coincides with the type graph language $\mathcal{L}(T)$.

Example 14. Let $\Lambda = \{A, B\}$ be given. The following is a duality pair:

$$(R, T) = \left(\underset{1}{\bullet} \xrightarrow{A} \underset{2}{\bullet} \xrightarrow{B} \underset{3}{\bullet} \; , \; A \circlearrowright \underset{1}{\bullet} \xleftarrow{A, B} \underset{2}{\bullet} \circlearrowleft B \right)$$

Since node 1 of T is not the source of a B-labeled edge and node 2 is not the target of an A-labeled edge, for every graph G we have $G \to T$ iff it does not contain a node which is both the target of an A-labeled edge and the source of a B-labeled edge. But it contains such a node if and only if $R \to G$.

One can identify the class of restriction graphs for which a corresponding type graph exists which defines the same graph language. Results from [15] state[3] that given a core graph R, a graph T can be constructed such that (R, T) is a duality pair if and only if R is a tree.

Thus we have a precise characterisation of the intersection of the classes of type and restriction graph languages: $\mathcal{L}$ belongs to the intersection if and only if it is of the form $\mathcal{L} = \mathcal{L}_R(R)$ and $core(R)$ is a tree. It is worth mentioning that the construction of T from R using the results from [15] contains two exponential blow-ups. This can be interpreted by saying that type graphs have limited expressiveness if used to forbid the presence of certain structures.

---

[2] Note that in [15] graphs are simple, but it can be easily seen that for our purposes the results can be transferred straightforwardly.

[3] We refer to Lemma 2.3, Lemma 2.5 and Theorem 3.1 in [15].

4   Type Graph Logic

In this section we investigate the possibility to define a language of graphs using a logical formula over type graphs. We start by defining the syntax and semantics of a type graph logic (*TGL*).

Definition 15 (Syntax and semantics of *TGL*). *A TGL formula F over a fixed set of edge labels Λ is formed according to the following grammar:*

$$F := T \mid F \vee F \mid F \wedge F \mid \neg F, \qquad where\ T\ is\ a\ type\ graph.$$

Each TGL formula F denotes a graph language $\mathcal{L}(F) \subseteq \mathbf{Gr}_\Lambda$ defined by structural induction as follows:

$$\mathcal{L}(T) = \{G \in \mathbf{Gr}_\Lambda \mid G \to T\} \qquad \mathcal{L}(\neg F) = \mathbf{Gr}_\Lambda \setminus \mathcal{L}(F)$$
$$\mathcal{L}(F_1 \wedge F_2) = \mathcal{L}(F_1) \cap \mathcal{L}(F_2) \qquad \mathcal{L}(F_1 \vee F_2) = \mathcal{L}(F_1) \cup \mathcal{L}(F_2)$$

Clearly, due to the presence of boolean connectives, boolean closure properties come for free.

Example 16. Let the following *TGL* formula F over $\Lambda = \{A, B\}$ be given:

$$F = \neg\ \bullet\!\!\bigcirc\!\!A \ \wedge\ \neg\ \bullet\!\!\bigcirc\!\!B$$

The graph language $\mathcal{L}(F)$ consists of all graphs which do not consist exclusively of A-edges or of B-edges, i.e., which contain at least one A-labeled edge and at least one B-labeled edge, something that can not be expressed by pure type graphs.

We now present some positive results for graph languages $\mathcal{L}(F)$ over *TGL* formulas F with respect to decidability problems. Due to the conjunction and negation operator, the emptiness (or unsatisfiability) check is not as trivial as it is for pure type graphs. Note that thanks to the presence of boolean connectives, inclusion can be reduced to emptiness.

Proposition 17. *For a graph language $\mathcal{L}(F)$ characterized by a TGL formula F, the following problems are decidable:*

- *Membership, i.e. for all graphs G it is decidable if $G \in \mathcal{L}(F)$ holds.*
- *Emptiness, i.e. it is decidable if $\mathcal{L}(F) = \varnothing$ holds.*
- *Language inclusion, i.e. given two TGL formulas F_1 and F_2 it is decidable if $\mathcal{L}(F_1) \subseteq \mathcal{L}(F_2)$ holds.*

Such a logic could alternatively also be defined based on restriction graphs. A related logic, for injective occurrences of restriction graphs, is studied in [17], where the authors also give a decidability result via inference rules.

5   Annotated Type Graphs

In this section we will improve the expressiveness of the type graphs themselves, rather than using an additional logic to do so. We will equip graphs with additional annotations. As explained in the introduction, this idea was already used similarly in abstract graph rewriting. In contrast to most other approaches, we will investigate the problem from a categorical point of view.

The idea we follow is to annotate each element of a type graph with pairs of multiplicities, denoting upper and lower bounds. We will define a category of multiply annotated graphs, where we consider elements of a lattice-ordered monoid (short ℓ-monoid) as multiplicities.

Definition 18 (Lattice-ordered monoid). *A lattice-ordered monoid (ℓ-monoid) $(\mathcal{M}, +, \leq)$ consists of a set $\mathcal{M}$, a partial order $\leq$ and a binary operation $+$ such that*

- *$(\mathcal{M}, \leq)$ is a lattice.*
- *$(\mathcal{M}, +)$ is a monoid; we denote its unit by 0.*
- *It holds that $a + (b \vee c) = (a + b) \vee (a + c)$ and $a + (b \wedge c) = (a + b) \wedge (a + c)$, where $\wedge, \vee$ are the meet and join of $\leq$.*

We denote by $\ell\mathbf{Mon}$ the category having ℓ-monoids as objects and as arrows monoid homomorphisms which are monotone.

Example 19. Let $n \in \mathbb{N} \backslash \{0\}$ and take $\mathcal{M}_n = \{0, 1, \ldots, n, m\}$ (zero, one, ..., n, many) with $0 \leq 1 \leq \cdots \leq n \leq m$ and addition as monoid operation with the proviso that $\ell_1 + \ell_2 = m$ if the sum is larger than n. Clearly, for all $a, b, c \in \mathcal{M}_n$ $a \vee b = \max\{a, b\}$ and $a \wedge b = \min\{a, b\}$. From this we can infer distributivity and therefore $(\mathcal{M}_n, +, \leq)$ forms an ℓ-monoid.

Furthermore, given a set S and an ℓ-monoid $(\mathcal{M}, +, \leq)$, it is easy to check that also $(\{a \colon S \to \mathcal{M}\}, +, \leq)$ is an ℓ-monoid, where the elements are functions from S to $\mathcal{M}$ and the partial order and the monoidal operation are taken pointwise.

In the following we will sometimes denote an ℓ-monoid by its underlying set.

Definition 20 (Annotations and multiplicities for graphs). *Given a functor $\mathcal{A} \colon \mathbf{Graph} \to \ell\mathbf{Mon}$, an annotation based on $\mathcal{A}$ for a graph G is an element $a \in \mathcal{A}(G)$. We write $\mathcal{A}_\varphi$, instead of $\mathcal{A}(\varphi)$, for the action of functor $\mathcal{A}$ on a graph morphism φ. We assume that for each graph G there is a standard annotation based on $\mathcal{A}$ that we denote by s_G, thus $s_G \in \mathcal{A}(G)$.*

Given an ℓ-monoid $\mathcal{M}_n = \{0, 1, \ldots, n, m\}$ we define the functor $\mathcal{B}^n \colon \mathbf{Graph} \to \ell\mathbf{Mon}$ as follows:

- *for every graph G, $\mathcal{B}^n(G) = \{a \colon (V_G \cup E_G) \to \mathcal{M}_n\}$;*
- *for every graph morphism $\varphi \colon G \to G'$ and $a \in \mathcal{B}^n(G)$, we have $\mathcal{B}^n_\varphi(a) \colon V_{G'} \cup E_{G'} \to \mathcal{M}_n$ with:*

$$\mathcal{B}^n_\varphi(a)(y) = \sum_{\varphi(x)=y} a(x), \quad \text{where } x \in (V_G \cup E_G) \text{ and } y \in (V_{G'} \cup E_{G'})$$

Therefore an annotation based on a functor $\mathcal{B}^n$ associates every item of a graph with a number (or the top value m). We will call such kind of annotations multiplicities. *Furthermore, the action of the functor on a morphism transforms a multiplicity by summing up (in $\mathcal{M}_n$) the values of all items of the source graph that are mapped to the same item of the target graph.*

For a graph G, its standard multiplicity $s_G \in \mathcal{B}^n(G)$ *is defined as the function which maps every node and edge of G to 1.*

Some of the results that we will present in the rest of the paper will hold for annotations based on a generic functor $\mathcal{A}$, some only for annotations based on functors $\mathcal{B}^n$, i.e. for multiplicities.

The type graphs which we are going to consider are enriched with a set of pairs of annotations. The motivation for considering multiple annotations rather than a single one is mainly to ensure closure under union. Each pair can be interpreted as establishing a lower and an upper bound to what a graph morphism can map to the graph.

Definition 21 (Multiply annotated graphs). *Given a functor $\mathcal{A}$: **Graph** → ℓ**Mon**, a multiply annotated graph $G[M]$ (over $\mathcal{A}$) is a graph G equipped with a finite set of pairs of annotations $M \subseteq \mathcal{A}(G) \times \mathcal{A}(G)$, such that $\ell \leq u$ for all $(\ell, u) \in M$.*

An arrow $\varphi \colon G[M] \to G'[M']$, also called a legal morphism, *is a graph morphism $\varphi \colon G \to G'$ such that for all $(\ell, u) \in M$ there exists $(\ell', u') \in M'$ with $\mathcal{A}_\varphi(\ell) \geq \ell'$ and $\mathcal{A}_\varphi(u) \leq u'$. We will write $G[\ell, u]$ as an abbreviation of $G[\{(\ell, u)\}]$. In case of annotations based on $\mathcal{B}^n$, we will often call a pair (ℓ, u) a* double multiplicity.

Multiply annotated graphs and legal morphisms form a category.

Lemma 22. *The composition of two legal morphisms is a legal morphism.*

Example 23. Consider the following multiply annotated graphs (over $\mathcal{B}^2$) $G[\ell, u]$ and $H[\ell', u']$, both having one double multiplicity.

$$G[\ell, u] = \underset{[1,1]}{\bullet} \xrightarrow{A\ [0,1]} \underset{[1,m]}{\bullet} \qquad\qquad H[\ell', u'] = \underset{[1,m]}{\bullet}\!\!\bigcirc A\ [0,m]$$

As evident from the picture, multiplicities are represented by writing the lower and upper bounds next to the corresponding graph elements. Note that there is a unique, obvious graph morphism $\varphi \colon G \to H$, mapping both nodes of G to the only node of H. Concerning multiplicities, by adding the lower and upper bounds of the two nodes of G, one gets the interval $[2, m]$ which is included in the interval of the node of H, $[1, m]$. Similarly, the double multiplicity $[0, 1]$ of the edge of G is included in $[0, m]$. Therefore, since both $\mathcal{B}^2_\varphi(\ell) \geq \ell'$ and $\mathcal{B}^2_\varphi(u) \leq u'$ hold, we can conclude that $\varphi \colon G[\ell, u] \to H[\ell', u']$ is a legal morphism.

We are now ready to define how a graph language $\mathcal{L}(T[M])$ looks like.

Definition 24. (Graph languages of multiply annotated type graphs).
We say that a graph G is represented by a multiply annotated type graph $T[M]$ whenever there exists a legal morphism $\varphi\colon G[s_G, s_G] \to T[M]$, i.e., there exists $(\ell, u) \in M$ such that $\ell \leq \mathcal{A}_\varphi(s_G) \leq u$. We will write $G \in \mathcal{L}(T[M])$ in this case. Whenever $M = \varnothing$ for a multiply annotated type graph $T[M]$ we get $\mathcal{L}(T[M]) = \varnothing$.

An extended example can be found in [6].

5.1   Decidability Properties for Multiply Annotated Graphs

We now address some decidability problems for languages defined by multiply annotated graphs. We get positive results with respect to the membership and emptiness problems. However, for decidability of language inclusion we only get partial results.

For the membership problem we can simply enumerate all graph morphisms $\varphi\colon G \to T$ and check if there exists a legal morphism $\varphi\colon G[s_G, s_G] \to T[M]$.

The emptiness check is somewhat more involved, since we have to take care of "illegal" annotations.

Proposition 25. *For a graph language $\mathcal{L}(T[M])$ characterized by a multiply annotated type graph $T[M]$ over $\mathcal{B}^n$ the emptiness problem is decidable: $\mathcal{L}(T[M]) = \varnothing$ iff $M = \varnothing$ or for each $(\ell, u) \in M$ there exists an edge $e \in E_T$ such that $\ell(e) \geq 1$ and $(u(src(e)) = 0$ or $u(tgt(e)) = 0)$.*

Language inclusion can be deduced from the existence of a legal morphism between the two multiply annotated type graphs.

Proposition 26. *The existence of a legal morphism $\varphi\colon T_1[M] \to T_2[N]$ implies $\mathcal{L}(T_1[M]) \subseteq \mathcal{L}(T_2[N])$.*

We would like to remark that this condition is sufficient but not necessary, and we present the following counterexample. Let the following two multiply annotated type graphs $T_1[M_1]$ and $T_2[M_2]$ over $\mathcal{B}^1$ be given where $|M_1| = |M_2| = 1$:

$$T_1[M_1] = \underset{[1,\, m]}{\bullet} \qquad\qquad T_2[M_2] = \underset{[1,\, 1]}{\bullet} \; \underset{[0,\, m]}{\bullet}$$

Clearly we have that the languages $\mathcal{L}(T_1[M_1])$ and $\mathcal{L}(T_2[M_2])$ are equal as both contain all discrete non-empty graphs. Thus $\mathcal{L}(T_1[M_1]) \subseteq \mathcal{L}(T_2[M_2])$, but there exists no legal morphism $\varphi\colon T_1[M_1] \to T_2[M_2]$. In fact, the upper bound of the first node of T_2 would be violated if the node of T_1 is mapped by φ to it, while the lower bound would be violated if the node of T_1 is mapped to the other node.

5.2   Deciding Language Inclusion for Annotated Type Graphs

In this section we show that if we allow only bounded graph languages consisting of graphs up to a fixed pathwidth, the language inclusion problem becomes decidable for annotations based on $\mathcal{B}^n$. Pathwidth is a well-known concept from graph theory that intuitively measures how much a graph resembles a path.

The proof is based on the notion of recognizability, which will be described via automaton functors that were introduced in [4]. We start with the main result and explain step by step the arguments that will lead to decidability.

Proposition 27. *The language inclusion problem is decidable for graph languages of bounded pathwidth characterized by multiply annotated type graphs over $\mathcal{B}^n$. That is, given $k \in \mathbb{N}$ and two multiply annotated type graphs $T_1[M_1]$ and $T_2[M_2]$ over $\mathcal{B}^n$, it is decidable whether $\mathcal{L}(T_1[M_1])^{\leq k} \subseteq \mathcal{L}(T_2[M_2])^{\leq k}$, where $\mathcal{L}(T[M])^{\leq k} = \{G \in \mathcal{L}(T[M]) \mid G \text{ has pathwidth} \leq k\}$.*

Our automaton model, given by automaton functors, reads cospans (i.e., graphs with interfaces) instead of single graphs. Therefore in the following, the category under consideration will be $Cospan_m(\mathbf{Graph})$, i.e. the category of cospans of graphs where the objects are discrete graphs J, K and the arrows are cospans $c: J \to G \leftarrow K$ where both graph morphisms are injective. We will refer to the graph J as the *inner interface* and to the graph K as the *outer interface* of the graph G. In addition we will sometimes abbreviate the cospan $c: J \to G \leftarrow K$ to the short representation $c: J \nrightarrow K$.

According to [3] a graph has pathwidth k iff it can be decomposed into cospans where each middle graph of a cospan has at most $k + 1$ nodes. Hence it is easy to check that a path has pathwidth 1, while a clique of order k has pathwidth $k - 1$.

Our main goal is to build an automaton which can read all graphs of our language step by step, similar to the idea of finite automata reading words in formal languages. Such an automaton can be constructed for an unbounded language, where the pathwidth is not restricted. However, we obtain a *finite* automaton only if we restrict the pathwidth. Then we can use well-known algorithms for finite automata to solve the language inclusion problem. Note that, if we would use tree automata instead of finite automata, our result could be generalized to graphs of bounded *treewidth*.

We will first introduce the notion of automaton functor (which is a categorical automaton model for so-called recognizable arrow languages) and which is inspired by Courcelle's theory of recognizable graph languages [9].

Definition 28 (Automaton functor [4]). *An automaton functor $\mathcal{C}$: $Cospan_m(\mathbf{Graph}) \to \mathbf{Rel}$ is a functor that maps every object J (i.e., every discrete graph) to a finite set $\mathcal{C}(J)$ (the set of states of J) and every cospan $c: J \nrightarrow K$ to a relation $\mathcal{C}(c) \subseteq \mathcal{C}(J) \times \mathcal{C}(K)$ (the transition relation of c). In addition there is a distinguished set of initial states $I \subseteq \mathcal{C}(\varnothing)$ and a distinguished set of final states $F \subseteq \mathcal{C}(\varnothing)$. The language $\mathcal{L}_\mathcal{C}$ of $\mathcal{C}$ is defined as follows:*

A graph G is contained in $\mathcal{L}_\mathcal{C}$ if and only if there exist states $q \in I$ and $q' \in F$ which are related by $\mathcal{C}(c)$, i.e. $(q, q') \in \mathcal{C}(c)$, where $c: \varnothing \to G \leftarrow \varnothing$ is the unique cospan with empty interfaces and middle graph G.

Languages accepted by automaton functors are called recognizable.

We will now define an automaton functor for a type graph $T[M]$ over $\mathcal{B}^n$.

Definition 29 (Counting cospan automaton). *Let $T[M]$ be a multiply annotated type graph over $\mathcal{B}^n$. We define an automaton functor $\mathcal{C}_{T[M]}$:* $\mathrm{Cospan}_m(\mathbf{Graph}) \to \mathbf{Rel}$ *as follows:*

- *For each object J of $\mathrm{Cospan}_m(\mathbf{Graph})$ (thus J is a finite discrete graph), $\mathcal{C}_{T[M]}(J) = \{(f, b) \mid f \colon J \to T, b \in \mathcal{B}^n(T)\}$ is its finite set of states*
- *$I \subseteq \mathcal{C}_{T[M]}(\varnothing)$ is the set of initial states with $I = \{(f \colon \varnothing \to T, 0)\}$, where 0 is the constant 0-function*
- *$F \subseteq \mathcal{C}_{T[M]}(\varnothing)$ is the set of final states with $F = \{(f \colon \varnothing \to T, b) \mid \exists (\ell, u) \in M : \ell \leq b \leq u\}$*
- *Let $c \colon J - \psi_L \to G \leftarrow \psi_R - K$ be an arrow in the category $\mathrm{Cospan}_m(\mathbf{Graph})$ with discrete interface graphs J and K where both graph morphisms $\psi_L \colon J \to G$ and $\psi_R \colon K \to G$ are injective. Two states $(f \colon J \to T, b)$ and $(f' \colon K \to T, b')$ are in the relation $\mathcal{C}_{T[M]}(c)$ if and only if there exists a morphism $h \colon G \to T$ such that the diagram below to the right commutes and for all $x \in V_T \cup E_T$ the following equation holds:*

$$b'(x) = b(x) + |\{y \in (G \setminus \psi_R(K)) \mid h(y) = x\}|$$

The set $G \setminus \psi_R(K)$ consists of all elements of G which are not targeted by the morphism ψ_R, e.g. $G \setminus \psi_R(K) = (V_G \setminus \psi_R(V_K)) \cup (E_G \setminus \psi_R(E_K))$. Instead of $\mathcal{L}_{\mathcal{C}_{T[M]}}$ and $\mathcal{C}_{T[M]}$ we just write $\mathcal{L}_C$ and $\mathcal{C}$ if $T[M]$ is clear from the context.

The intuition behind this construction is to count for each item x of T, step by step, the number of elements that are being mapped from a graph G (which is in the form of a cospan decomposition) to x, and then check if the bounds of a pair of annotations $(\ell, u) \in M$ of the multiply annotated type graph $T[M]$ are satisfied. We give a short example before moving on to the results.

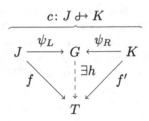

Example 30. Let the following multiply annotated type graph (over $\mathcal{B}^2$) $T[\ell, u]$ and the cospan $(c \colon \varnothing \to G \leftarrow \varnothing)$ with $G \in \mathcal{L}(T[\ell, u])$ be given:

$$T[\ell, u] = \quad \bullet \xrightarrow{\text{A } [0,2]} \bullet \circlearrowright \text{B } [0,m] \qquad c: \quad \varnothing \to \bullet \xrightarrow{\text{A}} \bullet \xrightarrow{\text{B}} \bullet \leftarrow \varnothing$$
$$\quad\quad [0,1] \quad\quad [1,m]$$

We will now decompose the cospan c into two cospans c_1, c_2 with $c = c_1 ; c_2$ in the following way:

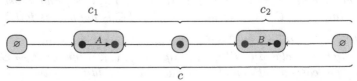

We let our counting cospan automaton parse the cospan decomposition $c_1; c_2$ step by step now to show how the annotations for the type graph T evolve during the process. According to our construction, every element in T has multiplicity 0 in the initial state of the automaton. We then sum up the number of elements within the middle graphs of the cospans which are *not* part of the right interface. Therefore we get the following parsing process:

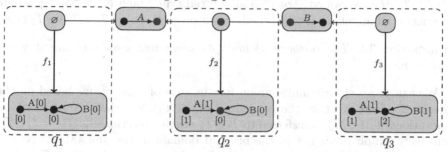

We visited three states q_1, q_2 and q_3 in the automaton with $(q_1, q_2) \in \mathcal{C}(c_1)$ and $(q_2, q_3) \in \mathcal{C}(c_2)$. Since $\mathcal{C}$ is supposed to be a functor we get that $\mathcal{C}(c_1); \mathcal{C}(c_2) = \mathcal{C}(c)$ and therefore $(q_1, q_3) \in \mathcal{C}(c)$ also holds. In addition we have $q_1 \in I$ and since the annotation function $b \in \mathcal{B}^2(T)$ in $q_3 = (f_3, b)$ satisfies $\ell \leq b \leq u$ we can infer that $q_3 \in F$. Therefore we can conclude that $G \in \mathcal{L}_\mathcal{C}$ holds as well.

We still need to prove that $\mathcal{C}$ is indeed a functor. Intuitively this shows that acceptance of a graph by the automaton is not dependent on its specific decomposition.

Proposition 31. *Let $c_1 \colon J \to G \leftarrow K$ and $c_2 \colon K \to H \leftarrow L$ be two arrows and let $id_G \colon G \to G \leftarrow G$ be the identity cospan.*

The mapping $\mathcal{C}_{T[M]} \colon \mathrm{Cospan}_m(\mathbf{Graph}) \to \mathbf{Rel}$ is a functor:

1. *$\mathcal{C}_{T[M]}(id_G) = id_{\mathcal{C}_{T[M]}(G)}$*
2. *$\mathcal{C}_{T[M]}(c_1; c_2) = \mathcal{C}_{T[M]}(c_1); \mathcal{C}_{T[M]}(c_2)$*

The language accepted by the automaton $\mathcal{L}_\mathcal{C}$ is exactly the graph language $\mathcal{L}(T[M])$.

Proposition 32. *Let the multiply annotated type graph $T[M]$ (over $\mathcal{B}^n$) and the corresponding automaton functor $\mathcal{C} \colon \mathrm{Cospan}_m(\mathbf{Graph}) \to \mathbf{Rel}$ for $T[M]$ be given. Then $\mathcal{L}_\mathcal{C} = \mathcal{L}(T[M])$ holds, i.e. for a graph G we have $G \in \mathcal{L}(T[M])$ if and only if there exist states $i \in I \subseteq \mathcal{C}(\varnothing)$ and $f \in F \subseteq \mathcal{C}(\varnothing)$ such that $(i, f) \in \mathcal{C}(c)$, where $c \colon \varnothing \to G \leftarrow \varnothing$.*

Therefore we can construct an automaton for each graph language specified by a multiply annotated type graph $T[M]$, which accepts exactly the same language. In case of a bounded graph language this automaton will have only finitely many states. Furthermore we can restrict the label alphabet, i.e., the cospans by using only atomic cospans, adding a single node or edges (see [2]). Once these steps are performed, we obtain conventional non-deterministic finite automata over a finite alphabet and we can use standard techniques from automata theory to solve the language inclusion problem directly on the finite automata.

5.3   Closure Properties for Multiply Annotated Graphs

Extending the expressiveness of the type graphs by adding multiplicities gives us positive results in case of closure under union and intersection. Here we use constructions that rely on products and coproducts in the category of graphs. Closure under intersection holds for the most general form of annotations. From $T_1[M_1]$, $T_2[M_2]$ we can construct an annotated type graph $(T_1 \times T_2)[N]$, where N contains all annotations which make both projections $\pi_i \colon T_1 \times T_2 \to T_i$ legal.

Proposition 33. *The category of multiply annotated graphs is closed under intersection.*

We can prove closure under union for the case of annotations based on the functor $\mathcal{B}^n$. Here we take the coproduct $(T_1 \oplus T_2)[N]$, where N contains all annotations of M_1, M_2, transferred to $T_1 \oplus T_2$ via the injections $i_j \colon T_j \to T_1 \oplus T_2$. Intuitively, graph items not in the original domain of the annotations receive annotation $[0, 0]$. This can be generalized under some mild assumptions (see proof in [6]).

Proposition 34. *The category of multiply annotated graphs over functor $\mathcal{B}^n$ is closed under union.*

Closure under complement is still an open issue. If we restrict to graphs of bounded pathwidth, we have a (non-deterministic) automaton (functor), as described in Sect. 5.1, which could be determinized and complemented. However, this does not provide us with an annotated type graph for the complement. We conjecture that closure under complement does not hold.

6   Conclusion

Our results on decidability and closure properties for specification languages are summarized in the following table. In the case where the results hold only for bounded pathwidth, the checkmark is in brackets.

		Pure TG	Restr. Gr.	TG Logic	Annotated TG
Decidability	$G \in \mathcal{L}$?	✓	✓	✓	✓
	$\mathcal{L} = \varnothing$?	✓	✓	✓	✓
	$\mathcal{L}_1 \subseteq \mathcal{L}_2$?	✓	✓	✓	(✓)
Closure Properties	$\mathcal{L}_1 \cup \mathcal{L}_2$	✗	✓	✓	✓
	$\mathcal{L}_1 \cap \mathcal{L}_2$	✓	✗	✓	✓
	$\mathbf{Gr}_\Lambda \setminus \mathcal{L}$	✗	✗	✓	?

One open question that remains is whether language inclusion for annotated type graphs is decidable if we do not restrict to bounded treewidth. Similarly, closure under complement is still open.

Furthermore, in order to be able to use these formalisms extensively in applications, it is necessary to provide a mechanism to compute weakest preconditions and strongest postconditions. This does not seem feasible for pure type graphs or the type graph logic. Hence, we are currently working on characterizing weakest preconditions and strongest postconditions in the setting of annotated type graphs. This requires a materialisation construction, similar to [23], which we plan to characterize abstractly, exploiting universal properties in category theory.

Note that our annotations are global, i.e., we count *all* items that are mapped to a specific item in the type graph. This holds also for edges, as opposed to UML multiplicities, which are local wrt. the classes which are related by an edge (i.e., an association). We plan to study the possibility to integrate this into our framework and investigate the corresponding decidability and closure properties.

Related work: As already mentioned there are many approaches for specifying graph languages. One can not say that one is superior to the other, usually there is a tradeoff between expressiveness and decidability properties, furthermore they differ in terms of closure properties.

Recognizable graph languages [8,9], which are the counterpart to regular word languages, are closely related with monadic second-order graph logic. If one restricts recognizable graph languages to bounded treewidth (or pathwidth as we did), one obtains satisfactory decidability properties. On the other hand, the size of the resulting graph automata is often quite intimidating [2] and hence they are difficult to work with in practical applications. The use of nested application conditions [13], equivalent to first-order logic [20], has a long tradition in graph rewriting and they can be used to compute pre- and postconditions for rules [18]. However, satisfiability and implication are undecidable for first-order logic.

A notion of grammars that is equivalent to context-free (word) grammars are hyperedge replacement grammars [12]. Many aspects of the theory of context-free languages can be transferred to the graph setting.

In heap analysis the representation of pointer structures to be analyzed requires methods to specify sets of graphs. Hence both the TVLA approach by Sagiv, Reps and Wilhelm [23], as well as separation logic [10,16] face this problem. In [23] heaps are represented by graphs, annotated with predicates from a three-valued logics (with truth values *yes*, *no* and *maybe*).

A further interesting approach are forest automata [1] that have many interesting properties, but are somewhat complex to handle.

In [22] the authors study an approach called Diagram Predicate Framework (DPF), in which type graphs have annotations based on generalized sketches. This formalism is intended for MOF-based modelling languages and allows more complex annotations than our framework.

References

1. Abdulla, P.A., Holík, L., Jonsson, B., Lengál, O., Trinh, C.Q., Vojnar, T.: Verification of heap manipulating programs with ordered data by extended forest automata. In: Hung, D., Ogawa, M. (eds.) ATVA 2013. LNCS, vol. 8172, pp. 224–239. Springer, Cham (2013). doi:10.1007/978-3-319-02444-8_17

2. Blume, C., Bruggink, H.J.S., Engelke, D., König, B.: Efficient symbolic implementation of graph automata with applications to invariant checking. In: Ehrig, H., Engels, G., Kreowski, H.-J., Rozenberg, G. (eds.) ICGT 2012. LNCS, vol. 7562, pp. 264–278. Springer, Heidelberg (2012). doi:10.1007/978-3-642-33654-6_18

3. Blume, C., Sander Bruggink, H.J., Friedrich, M., König, B.: Treewidth, pathwidth and cospan decompositions with applications to graph-accepting tree automata. J. Vis. Lang. Comput. **24**(3), 192–206 (2013)

4. Bruggink, H.J.S., König, B.: On the recognizability of arrow and graph languages. In: Ehrig, H., Heckel, R., Rozenberg, G., Taentzer, G. (eds.) ICGT 2008. LNCS, vol. 5214, pp. 336–350. Springer, Heidelberg (2008). doi:10.1007/978-3-540-87405-8_23

5. Clarke, E.M., Grumberg, O., Jha, S., Yuan, L., Veith, H.: Counterexample-guided abstraction refinement for symbolic model checking. J. ACM **50**(5), 752–794 (2003)

6. Corradini, A., König, B., Nolte, D.: Specifying graph languages with type graphs, arXiv:1704.05263 (2017)

7. Corradini, A., Montanari, U., Rossi, F.: Graph processes. Fundamenta Informaticae **26**(3/4), 241–265 (1996)

8. Courcelle, B.: The monadic second-order logic of graphs I. Recognizable sets of finite graphs. Inf. Comput. **85**, 12–75 (1990)

9. Courcelle, B., Engelfriet, J.: Graph Structure and Monadic Second-Order Logic, A Language-Theoretic Approach. Cambridge University Press, New York (2012)

10. Distefano, D., O'Hearn, P.W., Yang, H.: A local shape analysis based on separation logic. In: Hermanns, H., Palsberg, J. (eds.) TACAS 2006. LNCS, vol. 3920, pp. 287–302. Springer, Heidelberg (2006). doi:10.1007/11691372_19

11. Endrullis, J., Zantema, H.: Proving non-termination by finite automata. In: RTA 2015, vol. 36. LIPIcs, pp. 160–176. Schloss Dagstuhl-Leibniz-Zentrum fuer Informatik (2015)

12. Habel, A.: Hyperedge Replacement: Grammars and Languages. LNCS, vol. 643. Springer, Heidelberg (1992). doi:10.1007/BFb0013875

13. Habel, A., Pennemann, K.-H.: Nested constraints and application conditions for high-level structures. In: Kreowski, H.-J., Montanari, U., Orejas, F., Rozenberg, G., Taentzer, G. (eds.) Formal Methods in Software and Systems Modeling. LNCS, vol. 3393, pp. 293–308. Springer, Heidelberg (2005). doi:10.1007/978-3-540-31847-7_17

14. Heckel, R., Wagner, A.: Ensuring consistency of conditional graph rewriting - a constructive approach. In: Proceedings of the Joint COMPUGRAPH/SEMAGRAPH Workshop on Graph Rewriting and Computation, vol. 2, ENTCS (1995)

15. Nešetřil, J., Tardif, C.: Duality theorems for finite structures (characterising gaps and good characterisations). J. Comb. Theory Ser. B **80**, 80–97 (2000)

16. O'Hearn, P.W.: Resources, concurrency and local reasoning. Theor. Comput. Sci. **375**(1–3), 271–307 (2007). Reynolds Festschrift

17. Orejas, F., Ehrig, H., Prange, U.: A logic of graph constraints. In: Fiadeiro, J.L., Inverardi, P. (eds.) FASE 2008. LNCS, vol. 4961, pp. 179–198. Springer, Heidelberg (2008). doi:10.1007/978-3-540-78743-3_14

18. Pennemann, K.-H.: Development of Correct Graph Transformation Systems. Ph.D. thesis, Universität Oldenburg, May 2009

19. Rensink, A.: Canonical graph shapes. In: Schmidt, D. (ed.) ESOP 2004. LNCS, vol. 2986, pp. 401–415. Springer, Heidelberg (2004). doi:10.1007/978-3-540-24725-8_28

20. Rensink, A.: Representing first-order logic using graphs. In: Ehrig, H., Engels, G., Parisi-Presicce, F., Rozenberg, G. (eds.) ICGT 2004. LNCS, vol. 3256, pp. 319–335. Springer, Heidelberg (2004). doi:10.1007/978-3-540-30203-2_23

21. Rozenberg, G., (ed.): Handbook of Graph Grammars and Computing by Graph Transformation, vol. 1: Foundations. World Scientific (1997)

22. Rutle, A., Rossini, A., Lamo, Y., Wolter, U.: A diagrammatic formalisation of MOF-based modelling languages. In: Oriol, M., Meyer, B. (eds.) TOOLS EUROPE 2009. LNBIP, vol. 33, pp. 37–56. Springer, Heidelberg (2009). doi:10. 1007/978-3-642-02571-6_4
23. Sagiv, M., Reps, T., Wilhelm, R.: Parametric shape analysis via 3-valued logic. TOPLAS (ACM Trans. Program. Lang. Syst.) **24**(3), 217–298 (2002)
24. Steenken, D., Wehrheim, H., Wonisch, D.: Sound and complete abstract graph transformation. In: Simao, A., Morgan, C. (eds.) SBMF 2011. LNCS, vol. 7021, pp. 92–107. Springer, Heidelberg (2011). doi:10.1007/978-3-642-25032-3_7

Fusion Grammars: A Novel Approach to the Generation of Graph Languages

Hans-Jörg Kreowski, Sabine Kuske, and Aaron Lye[✉]

Department of Computer Science, University of Bremen,
P.O.Box 33 04 40, 28334 Bremen, Germany
{kreo,kuske,lye}@informatik.uni-bremen.de

Abstract. In this paper, we introduce the notion of fusion grammars as a novel device for the generation of (hyper)graph languages. Fusion grammars are motivated by the observation that many large and complex structures can be seen as compositions of a large number of small basic pieces. A fusion grammar is a hypergraph grammar that provides the small pieces as connected components of the start hypergraph. To get arbitrary large numbers of them, they can be copied multiple times. To get large connected hypergraphs, they can be fused by the application of fusion rules. As the first main results, we show that fusion grammars can simulate hyperedge replacement grammars that generate connected hypergraphs, that the membership problem is decidable, and that fusion grammars are more powerful than hyperedge replacement grammars.

1 Introduction

One encounters various fusion processes in various scientific fields like DNA computing, chemistry, tiling, fractal geometry, visual modeling and others. For example, the fusion of DNA double strands according to the Watson-Crick complemtarity is a key operation of DNA computing in the wake of the Adleman experiment (see, e.g. [1,2]). As an illustration, consider the DNA double strands

$$\overline{ATTA} \qquad \overline{GTA} \qquad \overline{C}$$
$$TA \qquad\qquad ATC \qquad\qquad ATG$$

As they have complementary sticky ends, the first strand and the second or the third one as well as the second and the third one can fuse with each other. Moreover, the second strand can fuse with itself. Therefore, in a tube with many identical molecules of the three kinds, one gets fused molecules like those given in Fig. 1a. Similar effects can be seen in the iteration of some fractals (see, e.g., [3]) like those depicted in Fig. 1b. This iterates the Sierpinski triangle where the $(i + 1)$-th structure is obtained from the i-th structure by zooming it with the factor $1/2$, making three copies and fusing the right corner of one copy with the left corner of another copy and the upper corners of both copies with the left and right corner of the third copy. Ignoring the geometry, the i-th iterated structure of the Sierpinski triangle is a fusion of 3^i copies of the initial triangle. Quite similar

© Springer International Publishing AG 2017
J. de Lara and D. Plump (Eds.): ICGT 2017, LNCS 10373, pp. 90–105, 2017.
DOI: 10.1007/978-3-319-61470-0_6

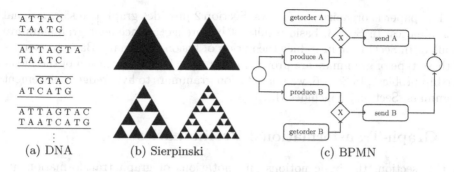

(a) DNA   (b) Sierpinski   (c) BPMN

Fig. 1. Examples of fused entities

fusion operations can be found in mosaics and tilings (see, e.g., [4]). As a last example, we may look into the area of visual modeling where one encounters quite a spectrum of diagrams that are composed of some basic forms. For instance, a BPMN process may look as depicted in Fig. 1c. In all the examples, the same principle works: A few small entities may be copied and fused to produces more complicated entities.

In this paper, we capture the principle in the formal framework of fusion grammars which are generative devices on hypergraphs. The basic entities are provided by the start hypergraph. The fusion is done by means of fusion rules that take two complemtary hyperedges with the same number of attachment nodes. By rule application, each two corresponding nodes of the attachment nodes are fused with each other while the hyperedges are consumed. This allows in particular to connect disconnected component hypergraphs. Moreover, arbitrary multiplication of connected components is admitted during the derivation process. Because of the multiplication, the derived hypergraphs may consist of many disjoint connected components that may not always be used for fusion in a meaningful way later on. Hence, we consider hypergraphs as a kind of "tubes of molecules" where the molecules are represented by the connected components. In this respect, fusion grammars are closely related to graph multiset transformation as one encounters in [5,6]. Consequently, the generated language contains terminal connected components of derived hypergraphs rather than the whole terminal derived hypergraphs. Additionally, we allow to use a marker mechanism to identify the language elements. Fusion grammars exhibit quite strong parallelism properties and it turns out that fusions and multiplications can be interchanged. Based on these results, one can prove that the membership problem of fusion grammars is decidable provided that none of the connected components of the start hypergraph consists of fusion hyperedges only. A second main result is that the well-known hyperedge replacement grammars can be transformed into fusion grammars such that the source grammar and the target grammar generate the same language if the source grammar has rules with connected right-hand sides. As a last result we show that fusion grammars are more powerful than hyperedge replacement grammars.

The paper is organized as follows. Section 2 provides graph-transformational preliminaries. In Sect. 3, basic results of hyperedge replacement grammars are recalled. In Sect. 4, we introduce the notion of fusion grammars. Basic properties of these type of grammars are proven in Sect. 5 including a solution of the membership problem. In Sect. 6, we relate fusion grammars to hyperedge replacement grammars. Section 7 concludes the paper.

2   Graph-Transformational Preliminaries

In this section, the basic notions and notations of graph transformation are recalled as far as needed (see, e.g., [7]). Although we are mainly interested in graph languages, we use the more general hypergraphs as underlying structures because they make some technicalities a bit simpler. We consider hypergraphs the hyperedges of which are attached to a sequence of nodes and labeled in a given label alphabet Σ.

A *hypergraph* over Σ is a system $H = (V, E, att, lab)$ where V is a finite set of *nodes*, E is a finite set of *hyperedges*, $att\colon E \to V^*$ is a function, called *attachment*, and $lab\colon E \to \Sigma$ is a function, called *labeling*.

The length of the attachment $att(e)$ for $e \in E$ is called *type* of e, and e is called A-hyperedge if A is its label. The components of $H = (V, E, att, lab)$ may also be denoted by V_H, E_H, att_H, and lab_H respectively. The class of all hypergraphs over Σ is denoted by $\mathcal{H}_\Sigma$.

A *graph* is a hypergraph $H = (V, E, att, lab)$ with $att(e) \in V^2$ for all $e \in E$. In this case, the hyperedges are called *edges*. Moreover, the attachment may be separated in *source* and *target* mappings $s, t\colon E \to V$ given by $s(e) = v$ and $t(e) = v'$ for $att(e) = vv'$. If $s(e) = t(e)$, then e is also called a *loop*.

In drawings, a hyperedge e with attachment $att(e) = v_1 \cdots v_k$ is depicted by

i.e. numbered tentacles connect the label with the corresponding attachment nodes. Moreover, an edge e is depicted by $\bullet\!\xrightarrow{A}\!\bullet$ instead of $\bullet\!\xrightarrow{\;1\;}\!A\!\xrightarrow{\;2\;}\!\bullet$ and $\overleftarrow{\bigcirc} A$ if $s(e) = t(e)$. We assume the existence of a special label $* \in \Sigma$ that is omitted in drawings. In this way, unlabeled hyperedges are represented by hyperedges labeled with $*$. If there are two edges with the same label, but in opposite directions, we may draw them as an undirected edge: $\bullet\!\overset{A}{\rule{1.5em}{0.4pt}}\!\bullet$ instead of $\bullet\!\overset{A}{\underset{A}{\rightleftarrows}}\!\bullet$.

Given $H, H' \in \mathcal{H}_\Sigma$, H is a *subhypergraph* of H' if $V_H \subseteq V_{H'}$, $E_H \subseteq E_{H'}$, $att_H(e) = att_{H'}(e)$, and $lab_H(e) = lab_{H'}(e)$ for all $e \in E_H$. This is denoted by $H \subseteq H'$.

Given $H, H' \in \mathcal{H}_\Sigma$, a *(hypergraph) morphism* $g\colon H \to H'$ consists of two mappings $g_V\colon V_H \to V_{H'}$ and $g_E\colon E_H \to E_{H'}$ such that $att_{H'}(g_E(e)) = g_V^*(att_H(e))$ and $lab_{H'}(g_E(e)) = lab_H(e)$ for all $e \in E_H$ where $g_V^*\colon V_H^* \to V_{H'}^*$ is the canonical extension of g_V, given by $g_V^*(v_1 \cdots v_n) = g_V(v_1) \cdots g_V(v_n)$ for all $v_1 \cdots v_n \in V_H^*$. H and H' are *isomorphic*, denoted by $H \cong H'$, if there is an isomorphism $g \colon H \to H'$, i.e. a morphism with bijective mappings. Clearly, $H \subseteq H'$ implies that the two inclusions $V_H \subseteq V_{H'}$ and $E_H \subseteq E_{H'}$ define a morphism from $H \to H'$. Given a morphism $g \colon H \to H'$, the image of H in H' under g defines a subgraph $g(H) \subseteq H'$.

Let $H' \in \mathcal{H}_\Sigma$ as well as $V \subseteq V_{H'}$ and $E \subseteq E_{H'}$. Then the *removal* of (V, E) from H' given by $H = H' - (V, E) = (V_{H'} - V, E_{H'} - E, att_H, lab_H)$ with $att_H(e) = att_{H'}(e)$ and $lab_H(e) = lab_{H'}(e)$ for all $e \in E_{H'} - E$ defines a subgraph $H \subseteq H'$ if $att_{H'}(e) \in (V_{H'} - V)^*$ for all $e \in E_{H'} - E$, i.e. no remaining hyperedge is attached to a node that does not remain. This condition is called *dangling condition* because the removal of a node would leave some dangling tentacles if this condition is violated.

Let $H \in \mathcal{H}_\Sigma$ and $H' = (V', E', att' \colon E' \to (V_H + V')^*, lab' \colon E' \to \Sigma)$ be some quadruple with two sets V', E' and two mappings att' and lab' where $+$ denotes the disjoint union of sets. Then the *extension* of H by H' given by $H + H' = (V_H + V', E_H + E', att, lab)$ with $att(e) = att_H(e)$ and $lab(e) = lab_H(e)$ for all $e \in E_H$ as well as $att(e) = att'(e)$ and $lab(e) = lab'(e)$ for all $e \in E'$ is a hypergraph with $H \subseteq H + H'$. A special case of extension is used several times further on, i.e. the addition of a single hyperedge with the attachment $1 \cdots k$ and label A. This extension is denoted by $([k] \subseteq H)^A$ with $[k] = \{1, \ldots, k\}$. If k is known from the context, then the shorter denotation H^A is used. If, in addition, H is the discrete graph with the nodes $1, \ldots, k$, also denoted by $[k]$, then one writes $A^\bullet$ instead of $[k]^A$. $A^\bullet$ is called a *handle*. Likewise, we write $(A, B)^\bullet$ if there are two hyperedges with labels A and B, the same attachment $1 \cdots k$ and no further nodes. The construction also works if $H' \in \mathcal{H}_\Sigma$. The only difference is that $att_{H'} \colon E_{H'} \to V_{H'}^*$ and $att' \colon E' \to (V_H + V')^*$ have different domains. But this causes no problem because $V_{H'}^* \subseteq (V_H + V_{H'})^*$. In this case, $H + H'$ is called *disjoint union* of H and H'.

As the disjoint union of sets is unique up to bijection, extensions of hypergraphs and disjoint unions of hypergraphs are also unique up to isomorphism. It is easy to see that the disjoint union is commutative and associative. Moreover, there are injections (injective morphisms) $in_H \colon H \to H + H'$ and $in_{H'} \colon H' \to H + H'$ such that $in_H(H) \cup in_{H'}(H') = H + H'$ and $in_H(H) \cap in_{H'}(H') = \emptyset$. Each two morphisms $g_H \colon H \to Y$ and $g_{H'} \colon H' \to Y$ define a unique morphism $\langle g_H, g_{H'} \rangle \colon H + H' \to Y$ with $\langle g_H, g_{H'} \rangle \circ in_H = g_H$ and $\langle g_H, g_{H'} \rangle \circ in_{H'} = g_{H'}$. In particular, one gets $g = \langle g \circ in_H, g \circ in_{H'} \rangle$ for all morphisms $g \colon H + H' \to Y$ and $g + g' = \langle in_Y \circ g, in_{Y'} \circ g' \rangle \colon H + H' \to Y + Y'$ for morphisms $g \colon H \to Y$ and $g' \colon H' \to Y'$. A special case is the disjoint union of H with itself k times, denoted by $k \cdot H$. We make frequently use of these facts throughout the paper.

The fusion of nodes is a further basic construction used in the paper. It is defined as a quotient by means of an equivalence relation $\equiv$ on the set of nodes V_H of $H \in \mathcal{H}_\Sigma$ as follows: $H/\equiv = (V_H/\equiv, E_H, att_{H/\equiv}, lab_H)$ with $att_{H/\equiv}(e) = [v_1] \cdots [v_k]$ for $e \in E_H$, $att_H(e) = v_1 \cdots v_k$ where $[v]$ denotes the equivalence class of $v \in V_H$ and $V_H/\equiv$ is the set of equivalence classes. It is easy to see that

$f\colon H \to H/\equiv$ given by $f_V(v) = [v]$ for all $v \in V_H$ and $f_E(e) = e$ for all $e \in E_H$ defines a *quotient morphism*.

Let $H \in \mathcal{H}_\Sigma$. Then a sequence of triples $(i_1, e_1, o_1) \ldots (i_n, e_n, o_n) \in (\mathbb{N} \times E_H \times \mathbb{N})^*$ is a *path* from $v \in V_H$ to $v' \in V_H$ if $v = att_H(e_1)_{i_1}$, $v' = att_H(e_n)_{o_n}$ and $att_H(e_j)_{o_j} = att_H(e_{j+1})_{i_{j+1}}$ for $j = 1, \ldots, n-1$ where, for each $e \in E_H$, $att_H(e)_i = v_i$ for $att_H(e) = v_1 \cdots v_k$ and $i = 1, \ldots, k$. H is *connected* if each two nodes are connected by a path. A subgraph C of H is a *connected component* of H if it is connected and there is no larger connected subgraph, i.e. $C \subseteq C' \subseteq H$ and C' connected implies $C = C'$. The set of connected components of H is denoted by $\mathcal{C}(H)$.

We use the *multiplication* of H defined by means of $\mathcal{C}(H)$ as follows. Let $m\colon \mathcal{C}(H) \to \mathbb{N}_{>0}$ be a mapping, called *multiplicity*, then $m \cdot H = \sum\limits_{C \in \mathcal{C}(H)} m(C) \cdot C$.

A *rule* $r = (L \supseteq K \to R)$ consists of $L, K, R \in \mathcal{H}_\Sigma$ with $K \subseteq L$ and a morphism $b\colon K \to R$ that is injective on E_K. To apply r to $H \in \mathcal{H}_\Sigma$, one needs first a *matching* morphism $g\colon L \to H$ subject to the *gluing condition* consisting of the dangling condition of $H - (g(L) - g(K))$, and the *identification condition*, i.e. $g_V(v) = g_V(v')$ implies $v = v'$ or $v, v' \in V_K$ for all $v, v' \in V_L$ and $g_E(e) = g_E(e')$ implies $e = e'$ or $e, e' \in E_K$ for all $e, e' \in E_L$. Then one removes $g(L) - g(K)$ from H yielding the intermediate hypergraph X with $X \subseteq H$ and a morphism $d\colon K \to X$ restricting g to K and X. Finally, X and R are merged along the morphisms d and b, i.e. the resulting hypergraph is $H' = (X + R)/(d = b)$ being the quotient of $X + R$ through the equivalence relation $d = b$ given by the relation $\{(d_V(v), b_V(v)) \mid v \in V_K\}$. The inclusions in_X and in_R into $X + R$ followed by the quotient morphism define morphisms $c\colon X \to H'$ and $h\colon R \to H'$ such that $c \circ d = h \circ b$ according to the definition of the equivalence $d = b$.

The application of a rule r to a hypergraph H is denoted by $H \underset{r}{\Longrightarrow} H'$ and called a *direct derivation*. Its sequential composition of direct derivations $der = (H = H_0 \underset{r_1}{\Longrightarrow} H_1 \underset{r_2}{\Longrightarrow} \cdots \underset{r_n}{\Longrightarrow} H_n = H')$ for some $n \in \mathbb{N}$ is called a *derivation* from H to H'. If $r_1, \ldots, r_n \in P$ (for some set P of rules), der can be denoted as $H \underset{P}{\overset{n}{\Longrightarrow}} H'$ or $H \underset{P}{\overset{*}{\Longrightarrow}} H'$ if the length does not matter.

Given two rules $r_i = (L_i \supseteq K_i \underset{b_i}{\longrightarrow} R_i)$ for $i = 1, 2$, the *parallel rule* of r_1 and r_2 is defined by $r_1 + r_2 = (L_1 + L_2 \supseteq K_1 + K_2 \underset{b_1+b_2}{\longrightarrow} R_1 + R_2)$. As parallel rules are rules, the construction of a direct derivation can be applied to parallel rules too. As the disjoint union is commutative and associative, one gets parallel rules $\sum_{i=1}^n r_i$ for each sequence $r_1 \cdots r_n$ of rules and each permutation of $r_1 \cdots r_n$ yields the same parallel rule. Given a set of rules, the set of all parallel rules is denoted by P_+.

Let $H \underset{r_1+r_2}{\Longrightarrow} H'$ be a direct derivation with the matching morphism $g\colon L_1 + L_2 \to H$. Then $g_i = g \circ in_{L_i}$ for $i = 1, 2$ yield the direct derivations $H \underset{r_i}{\Longrightarrow} H_i$ with the intermediate graphs X_i and the induced morphisms $c_i\colon X_i \to H_i$ for $i = 1, 2$. Moreover, the identification condition of g implies $g_1(L_1) = g(in_{L_1}(L_1)) \subseteq X_2$ and $g_2(L_2) = g(in_{L_2}(L_2)) \subseteq X_1$ such that g_1 and g_2 can be restricted to X_2 and X_1 respectively yielding the morphisms $e_1 \colon L_1 \to X_2$ and $e_2 \colon L_2 \to X_1$. This property is called *parallel independence*. Consequently, one gets matching morphisms $g_1' = c_2 \circ e_1$ and $g_2' = c_1 \circ e_2$ defining direct derivations $H_1 \underset{r_2}{\Longrightarrow} H_1'$ and $H_2 \underset{r_1}{\Longrightarrow} H_2'$. In [8], it is shown for directed labeled graphs that $H_1' \cong H' \cong H_2'$. The proof can be generalized to hypergraphs in a straightforward way. Besides this Sequentialization Theorem, there are two Parallelization Theorems. The first one states that two direct derivations $H \underset{r_i}{\Longrightarrow} H_i$ with the matching morphisms $g_i \colon L_i \to H$ for $i = 1, 2$ that are parallel independent induce a direct derivation $H \underset{r_1 + r_2}{\Longrightarrow} H'$ with the matching morphism $g = \langle g_1, g_2 \rangle \colon L_1 + L_2 \to H$. The second one concerns derivations of the form $H \underset{r_1}{\Longrightarrow} H_1 \underset{r_2}{\Longrightarrow} H'$. Such a derivation is called *sequentially independent* if there are morphisms $e_1' \colon R_1 \to X_2'$ restricting $h_1 \colon R_1 \to H_1$ and $e_2' \colon L_2 \to X_1$ restricting $g_2 \colon L_2 \to H_1$ where X_2' is the intermediate graph of $H_1 \underset{r_2}{\Longrightarrow} H'$ and X_1 is the intermediate graph of $H \underset{r_1}{\Longrightarrow} H_1$. In this case, one gets a parallel derivation step $H \underset{r_1 + r_2}{\Longrightarrow} H'$.

3 Hyperedge Replacement Grammars

In this section, we recall the notion of hyperedge replacement grammars (see, e.g., [9,10]) as it is a typical example of a language-generating device and as it is related to the novel notion of fusion grammars in Sect. 6.

Let $N \subseteq \Sigma$ be a set of nonterminals where each $A \in N$ has a *type* $k(A) \in \mathbb{N}$. Then a *hypergraph replacement rule* over N is of the form $A^\bullet \supseteq [k(A)] \subseteq R$ for some hypergraph R.

A *hyperedge replacement grammar* is a system $HRG = (N, T, P, S)$ where $N \subseteq \Sigma$ is a set of typed *nonterminals*, $T \subseteq \Sigma$ is a set of *terminals* with $T \cap N = \emptyset$, P is a set of hyperedge replacement rules over N and $S \in N$ is the *start symbol*. The language $L(HRG)$ is defined as $\{X \in \mathcal{H}_T \mid S^\bullet \underset{P}{\overset{*}{\Longrightarrow}} X\}$.

Example 1. Consider the hyperedge replacement grammar $SIER1 = (\{\triangle\}, \{*\}, \{\triangle^\bullet \supseteq [3] \subseteq R_1, \triangle^\bullet \supseteq [3] \subseteq R_2\}, \triangle)$ with R_1 and R_2 given as follows. The numbering of nodes indicates the inclusion of the three gluing nodes.

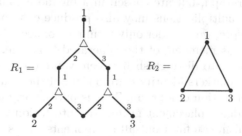

One may interpret the structures iterating the Sierpinski triangle sketched in the introduction as graphs using the corners of black triangles as nodes and their edges between corners as undirected edges. Then these graphs can be generated by $SIER1$ if one applies rule R_1 with maximum parallelism and then R_2 to terminate the generation. But as a hyperedge replacement grammar does not demand such a strict order of rule applications, $SIER1$ can also generate less regular structures.

Of all the known very nice properties of hyperedge replacement grammars, two are used in this paper.

Fact 1. *Let HRG be a hyperedge replacement grammar with connected right-hand sides. Then $L(HRG)$ contains only connected hypergraphs.*

Fact 2 (Context freeness lemma). *Let $HRG = (N, T, P, S)$ be a hyperedge replacement grammar. Let $A^\bullet \xRightarrow[P]{n} H$ be a derivation with $A \in N$ and $H \in \mathcal{H}_T$. Then there is a rule $A^\bullet \supseteq [k(A)] \subseteq R$, and for each $e \in E_R$ there is a hypergraph $H(e) \in \mathcal{H}_T$ such that $lab_R(e)^\bullet \xRightarrow{n(e)} H(e)$ with $\sum_{e \in E_R} n(e) = n - 1$ and $REPL(R, repl) = H$ where $repl: E_R \to \mathcal{H}_T$ is given by $repl(e) = H(e)$ and $REPL(R, repl)$ is the parallel application of the rules $lab_R(e)^\bullet \supseteq [k(lab_R(e))] \subseteq H(e)$ to $e^\bullet$, the subhypergraph of R consisting of e and its attachment.*

Note that the rule related to $e \in E_R$ is defined because hyperedge replacement does not remove nodes. We can assume $[k(A)] \subseteq H$ whenever $A^\bullet \xRightarrow{*} H$.

4   Fusion Grammars

In this section, we introduce the notion of *fusion grammars* as a novel type of graph grammars. Besides a start hypergraph and a specification of generated hypergraphs, a fusion grammar provides a set of fusion rules. The application of a fusion rule merges certain nodes which are given by two complementary hyperedges. Complementarity is defined on a set F of fusion labels that comes together with a complementary label $\overline{A}$ for each $A \in F$. Moreover, we assume that each $A \in F$ has a type $k(A) \in \mathbb{N}$ which is the number of tentacles of A-labeled hyperedges. Given a hypergraph, the set of all possible fusions is finite as fusion rules never create anything. To overcome this limitation, we allow arbitrary multiplications of disjoint components within derivations. The language generated by a fusion grammar does not consist of all terminal hypergraphs that are derived from the start hypergraph, but are chosen in a bit more complicated way. The problem is that the multiplications may also produce components that are not really needed. Therefore, we consider only terminal connected components of the derived hypergraphs as members of the generated language. Moreover, we use markers. They allow us to distinguish between wanted and unwanted terminal components; that is, markers identify components of the start hypergraph that contribute to the generation of a graph. This is particularly useful in the transformation of hyperedge replacement grammars into fusion grammars in Sect. 6 because this mechanism can filter out all components that stem from the initial handle of the hyperedge replacement grammar. Fusion grammars are illustrated by two examples where the second one uses a marker.

Definition 1 (Fusion rule). Let $F \subseteq \Sigma$ and $k: F \to \mathbb{N}$ be a type function. Let $\overline{A} \notin F$ be the complementary label for each $A \in F$ such that $\overline{A} \neq \overline{B}$ for all $A \neq B$. The typing is extended to complementary labels by $k(\overline{A}) = k(A)$ for all $A \in F$. Then $A \in F$ specifies the following *fusion rule*:

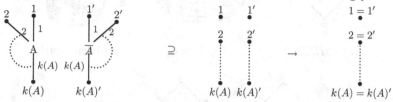

This rule is denoted by $fr(A)$. The underlying set F of labels is called *fusion alphabet*. The number at the nodes identify them so that the left-hand side inclusion and the right-hand side morphism are made visible. The morphism maps each attachment node and its primed counterpart to the same right-hand side node.

Definition 2 (Fusion grammar, derivation, and generated language). A *fusion grammar* is a system $FG = (Z, F, M, T)$ where Z is a *start hypergraph*, $F \subseteq \Sigma$ is a fusion alphabet, $M \subseteq \Sigma$ with $M \cap (F \cup \overline{F}) = \emptyset$ is a set of *markers*, and $T \subseteq \Sigma$ with $T \cap (F \cup \overline{F}) = \emptyset = T \cap M$ is a set of *terminal labels*.

A *derivation step* $H \Longrightarrow H'$ for some $H, H' \in \mathcal{H}_\Sigma$ is either a rule application $H \underset{r}{\Longrightarrow} H'$ for some parallel rule over $fr(F) = \{fr(f) \mid f \in F\}$ or a multiplication $H \underset{m}{\Longrightarrow} m \cdot H$ for some multiplicity m.

$L(FG) = \{rem_M(Y) \mid Z \overset{*}{\Longrightarrow} H, Y \in \mathcal{C}(H) \cap (\mathcal{H}_{T \cup M} - \mathcal{H}_T)\}$ is the *generated language* of FG where $rem_M(Y)$ is the hypergraph obtained when removing all hyperedges with labels in M from Y.

Remark 1. If all components of the start hypergraph have hyperedges with markers, then all connected components of derived hypergraphs have hyperedges with markers so that markers have no selective effect. This defines a special case of fusion grammars where the set of markers can be dropped and the hyperedges with markers removed from the start hypergraph. For such a fusion grammar FG, called *fusion grammar without markers*, the generated language is defined by $L(FG) = \{Y \mid Z \overset{*}{\Longrightarrow} H, Y \in \mathcal{C}(H) \cap \mathcal{H}_T\}$.

Example 2. Consider the set $\{N, W, S, E\}$ and let $F = \{N, W\}$ with $k(N) = k(W) = 1$ and $\overline{N} = S, \overline{W} = E$. Then the fusion grammar without markers $PSEUDOTORI = (CHECK, F, \{*\})$ with $CHECK$ depicted in Fig. 2a generates graphs of structures related to tori and Klein bottles, as the following reasoning indicates.

Starting with a multiplication by – say – 20, the fusion of disjoint components only yields structures like the ones depicted in Fig. 2d where the loops are omitted. If one continues now with fusions within connected components as long as possible, then one gets the resulting connected components as members of the generated language. Consider particularly the first connected component

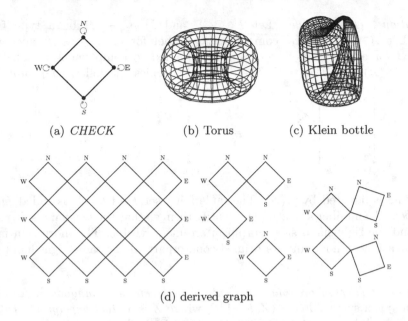

(a) *CHECK*   (b) Torus   (c) Klein bottle

(d) derived graph

Fig. 2. Pseudotori

in Fig. 2d. If one fuses the nodes with W- and E-loops from the top to bottom and the nodes with N- and S-loops from left to right, then one gets a torus (Fig. 2b, for simplicity the checks are replaced by rectangles). If one fuses the nodes with W- and E-loops in opposite order, then one gets a Moebius strip. The further fusion of the nodes with N- and S-loops from left to right yields the Klein bottle (Fig. 2c). All other terminal structures where the fusion is done in arbitrary order result in something between torus and Klein bottle. Hence we call them pseudotori.

Example 3. Consider the fusion grammar $SIER2 = ((\triangle, \#)^\bullet + R_1^\blacktriangle + R_2^\blacktriangle, \{\triangle\}, \{\#\}, \{*\})$ where $\triangle \in \Sigma$ is a fusion label with $\overline{\triangle} = \blacktriangle$ and $k(\triangle) = 3$. The three components of the start hypergraph are depicted in Fig. 3a: $(\triangle, \#)^\bullet$ in the upper left, $R_1^\blacktriangle$ on the right, and $R_2^\blacktriangle$ in the lower left.

To illustrate how the derivations work, one may start by a multiplication of $R_1^\blacktriangle$ by 2 and $R_2^\blacktriangle$ by 5. Then we may fuse the upper $\triangle$-hyperedge of one $R_1^\blacktriangle$ with the $\blacktriangle$-hyperedge of the other $R_1^\blacktriangle$ and all remaining five $\triangle$-hyperedges of the two fused $R_1^\blacktriangle$s with the $\blacktriangle$-hyperedges of the five $R_2^\blacktriangle$s. This results in the hypergraph in Fig. 3b. The fusion of its hyperedge with the $\triangle$-hyperedge of $(\triangle, \#)^\bullet$ yields the hypergraph in Fig. 3b – but with an #-hyperedge instead of the $\blacktriangle$-hyperedge – so that one gets a member of the generated language if one removes the #-hyperedge.

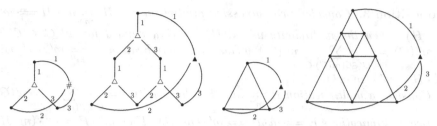

(a) The connected components of the start hypergraph   (b) Derived hypergraph

Fig. 3. The hypergraphs in $SIER2$

5   Properties of Fusion Grammars

In this section, we show that fusion grammars have very strong interchange and parallelization properties. The proofs of Propositions 1–3 are omitted because they follow straightforwardly from the definitions of fusion, multiplication and independence. The main result of this section states that the membership problem for fusion grammars is decidable if they are substantial.

Proposition 1. *Two fusions* $H \underset{fr(A)}{\Longrightarrow} H_1$ *and* $H \underset{fr(B)}{\Longrightarrow} H_2$ *are parallel independent if and only if their matches are hyperedge-disjoint. Two successive fusions* $H \underset{fr(A)}{\Longrightarrow} H_1 \underset{fr(B)}{\Longrightarrow} H'$ *are always sequentially independent.*

Note that two matches on the same hypergraph are always hyperedge-disjoint if $A \neq B$, in the case of $A = B$ one must make sure that the matches neither share an A-hyperedge nor an $\overline{A}$-hyperedge.

Proposition 2. *Consider a fusion* $C + H \underset{fr(A)}{\Longrightarrow} C_1 + H$ *for some connected component* C *followed by a multiplication* $C_1 + H \Longrightarrow k \cdot C_1 + H$. *Then fusion and multiplication can be exchanged in the following way:* $C + H \Longrightarrow k \cdot C + H \underset{k \cdot fr(A)}{\Longrightarrow} k \cdot C_1 + H$.

Analogously, a fusion $C + C' + H \underset{fr(A)}{\Longrightarrow} C_1 + H$ *for two connected components followed by a multiplication* $C_1 + H \Longrightarrow k \cdot C_1 + H$ *yields* $C + C' + H \Longrightarrow k \cdot C + k \cdot C' + H \underset{k \cdot fr(A)}{\Longrightarrow} k \cdot C_1 + H$.

Finally, consider a fusion $D + H \underset{fr(A)}{\Longrightarrow} C_1 + H$, *where* D *consists of one or two connected components and* $H = C_2 + H'$, *followed by a multiplication* $C_1 + H = C_1 + C_2 + H' \Longrightarrow C_1 + k \cdot C_2 + H'$. *Then one can also get:* $D + H = D + C_2 + H' \Longrightarrow D + k \cdot C_2 + H' \underset{fr(A)}{\Longrightarrow} C_1 + k \cdot C_2 + H'$.

Note that one must multiply the fusions in the first two cases to get the same result. So it may be reasonable to delay multiplications.

Proposition 3. *Consider two successive multiplications* $H \underset{m}{\Longrightarrow} m \cdot H \underset{m'}{\Longrightarrow} m' \cdot$
$(m \cdot H)$*. Then the multiplicity* $m'' \colon \mathcal{C}(H) \to \mathbb{N}_{>0}$ *defined for all* $C \in \mathcal{C}(H)$
by $m''(C) = \sum\limits_{C' \in \mathcal{C}(m(C) \cdot C)} m'(C')$ *yields the multiplication* $H \underset{m''}{\Longrightarrow} m'' \cdot H$ *with*
$m'' \cdot H = m' \cdot (m \cdot H)$.

    Consider a multiplication $H \underset{m''}{\Longrightarrow} m'' \cdot H$ *where* $m''(C) = \sum\limits_{C' \in \mathcal{C}(m(C) \cdot C)} m'(C')$
for two multiplications $H \underset{m}{\Longrightarrow} m \cdot H \underset{m'}{\Longrightarrow} m' \cdot (m \cdot H)$*. Then* $m'' \cdot H = m' \cdot (m \cdot H)$.

    In particular, $H \underset{m''}{\Longrightarrow} m'' \cdot H$ *with* $m''(C_0) \geq 2$ *for some* $C_0 \in \mathcal{C}(H)$ *can be*
decomposed in the following way. Let $m \colon \mathcal{C}(H) \to \mathbb{N}_{>0}$ *and* $m' \colon \mathcal{C}(m \cdot H) \to \mathbb{N}_{>0}$
be defined by $m(C_0) = m''(C_0)$*,* $m(C) = 1$ *for all* $C \in \mathcal{C}(H) - \{C_0\}$*,* $m'(C') = 1$
if $C' \in \mathcal{C}(m(C_0) \cdot C_0)$ *and* $m'(C') = m''(C')$ *otherwise. Then* $H \underset{m}{\Longrightarrow} m(C_0) \cdot C_0 +$
$(H - C_0) \underset{m'}{\Longrightarrow} m'' \cdot H$.

    Altogether, the results above yield the following corollary.

Corollary 1. *For each derivation* $H \overset{*}{\Longrightarrow} X$ *in a fusion grammar* FG*, one finds*
a fully sequentialized derivation $H \overset{*}{\Longrightarrow} X$ *consisting of single fusions and multi-*
plications of single connected components on one hand and a most parallelized
derivation $H \underset{m}{\Longrightarrow} m \cdot H \underset{fr(F)_+}{\Longrightarrow} X$ *on the other hand.*

    In order to generate hypergraphs with terminal hyperedges, the start hyper-
graph must contain at least one terminal hyperedge. Fusion grammars in which
every component of the start hypergraph contains terminal edges are called
substantial. In these grammars each derivation step increases the number of terminal hyperedges. For substantial fusion grammars the membership problem is
decidable.

Theorem 1. *Let* $FG = (Z, F, T)$ *or* $FG = (Z, F, M, T)$ *be a fusion grammar*
that is substantial, i.e. $C \notin \mathcal{H}_{F \cup \overline{F}}$ *for all* $C \in \mathcal{C}(Z)$*. Then the membership*
problem for $L(FG)$ *is decidable.*

Proof. If $FG = (Z, F, T)$ and $H \in L(FG)$, then there is a derivation $Z \overset{*}{\Longrightarrow} X$
with $H \in \mathcal{C}(X) \cap \mathcal{H}_T$. As FG is substantial, each $C \in \mathcal{C}(Z)$ contains at least
one terminal hyperedge. Hence, one needs at most the fusion of k connected
components of Z to get H where k is the number of hyperedges in H. In other
words, one may start with the multiplication $Z \underset{k}{\Longrightarrow} k \cdot Z$, and then try all possible fusions. The number of possible fusions is finite such that the procedure
terminates. If H appears in this way, then $H \in L(FG)$. Otherwise, $H \notin L(FG)$.
    In the case of $FG = (Z, F, M, T)$, a similar reasoning works. The only differ-
ence is that the fusion phase must involve, in addition to the connected components with terminal hyperedges, the connected components of Z that have no
terminal hyperedges, but marker hyperedges. If at all, a single fusion of this kind
is enough, to produce a hypergraph with marker. The process remains finite. If
H together with some marker hyperedge appears, then $H \in L(FG)$. Otherwise,
$H \notin L(FG)$.

Remark 2. As terminal and marker hyperedges do not get lost in fusions, all components of derived hypergraphs have terminal or marker hyperedges if the fusion grammar is substantial. Moreover, the result of a fusion has more terminal or marker hyperedges than each of the fusion components. This is the reason that membership can be decided in a similar way as for monotonic Chomsky grammars. Concering our examples, only *PSEUDOTORI* is substantial, but *SIER*2 is equivalent to the hyperedge replacement grammar *SIER*1 for which the membership problem is known to be decidable.

6 Transformation of Hyperedge Replacement Grammars into Fusion Grammars

The language of the fusion grammar *SIER*2 equals the language of the hyperedge replacement grammar *SIER*1. This is not a mere coincidence, but exemplifies the general relation between the two classes of grammars.

Each hyperedge replacement can be simulated by a fusion of the replaced hyperedge and the replacing right-hand side of the applied rule if one adds a hyperedge with complementary label to the right-hand side with the gluing nodes as attachment. This is the basic observation that leads to a transformation of hyperedge replacement grammars into fusion grammars. Formally, hypergraph representations of the hyperedge replacement rules are added disjointly to the initial handle of the hyperedge replacement grammars. To make sure that the constructed fusion grammars generate the same language as the given hyperedge replacement grammar, we must take into account that fusion grammars generate connected hypergraphs while hyperedge replacement grammars can also generate disconnected hypergraphs. Moreover, we use a marker to identify those fused hypergraphs that involve the initial handle.

Definition 3 (Hypergraph representation of hyperedge replacement rules). Let $N \subseteq \Sigma$ be a set of labels with a typing $k\colon N \to \mathbb{N}$ and a complementary label $\overline{A}$ for each $A \in N$ such that $\overline{A} \notin N$ and $\overline{A} \neq \overline{B}$ if $A \neq B$. Let $r = (A^\bullet \supseteq [k(A)] \subseteq R)$ be a hyperedge replacement rule. Then the *hyperedge representation* of the hyperedge replacement rule r is defined by $hgr(r) = R^{\overline{A}}$.

Proposition 4. *Using the notations of Definition 3, the following implication holds:* $H \underset{r}{\Longrightarrow} H'$ *implies* $H + hgr(r) \underset{fr(A)}{\Longrightarrow} H'$ *for all hypergraphs* H, H'.

Proof. Consider the graph H, depicted in Fig. 4a. Applying the rule r results in the graph depicted in Fig. 4c. Now consider H together with the hyperedge representation of r, depicted in Fig. 4b. A fusion rule $fr(A)$ fuses the two hyperedges with labels A and $\overline{A}$ such that the resulting graph is also the one depicted in Fig. 4c.

Definition 4 (From hyperedge replacement to fusion grammars). Let $HRG = (N, T, P, S)$ be a hyperedge replacement grammar. Let $\overline{A} \notin N$ be a complementary label for each $A \in N$ with $\overline{A} \neq \overline{B}$ if $A \neq B$. Then the corresponding

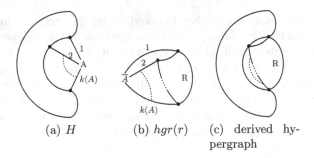

(a) H (b) $hgr(r)$ (c) derived hypergraph

Fig. 4. Components in the proof of Proposition 4

fusion grammar is defined by $FG(HRG) = ((S, \#)^{\bullet} + \sum_{r \in P} hgr(r), N, \{\#\}, T)$
with $\# \notin N \cup T$.

Example 4. Looking at the Examples 1 and 3, we have $FG(SIER1) = SIER2$.

Theorem 2. *Let $HRG = (N, T, P, S)$ be a hyperedge replacement grammar with connected right-hand side of rules. Then $L(HRG) = L(FG(HRG))$.*

Proof. We start to show that $H \in L(HRG)$ implies $H \in L(FG(HRG))$. H is terminal and connected as stated in Fact 1. Moreover, we can assume without loss of generality that $[k(S)] \subseteq H$. If $H^{\overline{S}}$ is derivable from $hgr(P) = \sum_{r \in P} hgr(r)$
then one can fuse $(S, \#)^{\bullet}$ with $H^{\overline{S}}$ yielding $H^{\#}$. Therefore, $H \in L(FG(HRG))$
after the removal of the marking $\#$-hyperedge. We prove the following more general statement by induction over the length of the derivation:

Claim. For all $A \in N$, for all $n \geq 1$, $A^{\bullet} \overset{n}{\Longrightarrow} H$ with $H \in \mathcal{H}_T$ implies $hgr(P) \overset{*}{\Longrightarrow} X$
with $H^{\overline{A}} \subseteq X$.

Induction base: $A^{\bullet} \overset{1}{\Longrightarrow} H$ means $r = (A^{\bullet} \supseteq [k(A)] \subseteq H) \in P$ and, therefore,
$hgr(P) \overset{0}{\Longrightarrow} hgr(P)$ and $hgr(r) = R^{\overline{A}} \subseteq hgr(P)$.

Induction step: Consider $A^{\bullet} \overset{n}{\Longrightarrow} H$ with $H \in \mathcal{H}_T$ and $n > 1$. Due to the Contextfreeness Lemma for hyperedge replacement grammars (cf. Fact 2), the derivation can be decomposed into its first step $A^{\bullet} \underset{r}{\Longrightarrow} R$ applying the rule $r = (A^{\bullet} \supseteq [k(A)] \subseteq R)$ and the remaining derivation $R \overset{n-1}{\Longrightarrow} H$ where the latter decomposes into derivations $lab_R(e)^{\bullet} \overset{n(e)}{\Longrightarrow} H(e)$ for all $e \in E_R$ such that $\sum_{e \in E_R} n(e) = n - 1$ and $REPL(E_R, repl) = H$ with $repl(e) = H(e)$
for $e \in E_R$. In particular, we have $n(e) \leq n - 1$ for $e \in E_R$ so that the induction hypothesis can be applied yielding, for each $e \in E_R$, a derivation $hgr(P) \overset{*}{\Longrightarrow} X(e)$ with $\overline{H(e)} = H(e)^{\overline{lab_R(e)}} \subseteq X(e)$. Then one can construct a derivation of the form $hgr(P) \Longrightarrow hgr(r) + k \cdot hgr(P) \overset{*}{\Longrightarrow} hgr(r) + \sum_{e \in E} X(e)$ for the rule r where k is the number of hyperedges of R. The hypergraph representation of the rule r is multiplied by $k + 1$, all the others by k. After

the multiplication, each of the components $hgr(P)$ is derived separately into $X(e)$. Now each $e \in E_R$ in $hgr(r)$ can be fused with the $\overline{lab_R(e)}$-hyperedge of $\overline{H}(e)$. Due to Proposition 4, this results in $\overline{H} = H^{\overline{A}}$ so that we have a derivation $hgr(P) \stackrel{*}{\Longrightarrow} hgr(r) + \sum_{e \in E_R} X(e) \stackrel{*}{\Longrightarrow} X = \overline{H} + \sum_{e \in E_R} (X(e) - \overline{H(e)})$ as stated. This completes the proof by induction. The statement applies to the derivation $S^{\bullet} \stackrel{*}{\Longrightarrow} H$ with $H \in \mathcal{H}_T$. Altogether, we get $H \in L(FG(HRG))$.

Conversely, we must show that $H \in L(FG(HRG))$ implies $H \in L(HRG)$. According to the definition of $FG(HRG), H \in \mathcal{H}_T$, and H is connected. Moreover, there is a derivation $(S, \#)^{\bullet} + hgr(P) \stackrel{*}{\Longrightarrow} X$ where X contains a connected component $H^{\#}$. The only way to derive a hypergraph with a $\#$-hyperedge is that, eventually, a fusion of the S-hyperedge of the start component $(S, \#)^{\bullet}$ with an $\overline{S}$-hyperedge takes place. Moreover by construction there is never a connected component with more than one $\#$-hyperedge. Without loss of generality, we can assume that this very fusion is the last one so that the derivation decomposes into $(S, \#)^{\bullet} + hgr(P) \stackrel{*}{\Longrightarrow} (S, \#)^{\bullet} + H^{\overline{S}} + X' \underset{fr(S)}{\Longrightarrow} X$ for some hypergraph X'. As no further hypergraph with a $\#$-hyperedge is needed, we can assume that $(S, \#)^{\bullet}$ is not multiplied during the derivation so that we can restrict the derivation to $hgr(P) \stackrel{*}{\Longrightarrow} H^{\overline{S}} + X'$ and must make sure that there is a derivation $S^{\bullet} \stackrel{*}{\Longrightarrow} H$ in HRG. We prove the following more general statement by induction:

Claim. Let $hgr(P) \stackrel{n}{\Longrightarrow} Y$ and $\overline{H}$ be a connected component of Y with a complementary hyperedge. Then $\overline{H} = H^{\overline{A}}$ for some $A \in N$ and $A^{\bullet} \stackrel{*}{\Longrightarrow} H$.

Induction base: $hgr(P) \stackrel{0}{\Longrightarrow} hgr(P)$. The connected components are $R^{\overline{A}}$ for all rules so that one also gets $A^{\bullet} \Longrightarrow R$.

Induction step: Consider $hgr(P) \stackrel{n+1}{\Longrightarrow} Y$ for some $n \in \mathbb{N}$. The derivation can be decomposed into $hgr(P) \stackrel{n}{\Longrightarrow} Y' \Longrightarrow Y$ where, without loss of generality, the last step is a multiplication or a single fusion. Due to the induction hypothesis, the connected components of Y' have the stated properties. If the last step is a multiplication, then the multiples in Y are isomorphic to their originals in Y' so that the statement holds for the connected components of Y, too. If the last step is a fusion, without loss of generality, an application of the fusion rule $fr(B)$ to a connected component $\overline{H}$ (Case 1) or to two connected components $\overline{H}', \overline{H}''$ (Case 2). All other connected components of Y stem from Y' so that they have the claimed property by induction hypothesis. As the connected components of the start hypergraph have at most one complementary hyperedge and as a complementary hyperedge is consumed by each fusion, the connected components of derived hypergraphs have also at most one complementary hyperedge. Therefore, the result of the fusion in Case 1 has no complementary hyperedge so that the claim holds trivially. In Case 2, we know by induction hypothesis that $\overline{H}' = H'^{\overline{A}}$ and $\overline{H}'' = H''^{\overline{C}}$ for some $A, C \in N$ as well as that there are $A^{\bullet} \stackrel{*}{\Longrightarrow} H'$ and $C^{\bullet} \stackrel{*}{\Longrightarrow} H''$. Because of the applicability of $fr(B)$, we know that H' has a B-hyperedge and $B = C$. The fusion of $H'^{\overline{A}}$ and $H''^{\overline{B}}$ yields a hypergraph $H^{\overline{A}}$ for some H. Moreover, one can apply the hyperedge replacement rule

$B^\bullet \supseteq [k(B)] \subseteq H''$ to H' yielding $A^\bullet \overset{*}{\Longrightarrow} \hat{H}$. According to Proposition 4, we get $H = \hat{H}$ so that the claim holds also in Case 2. This completes the proof.

If one can embed a class of languages into another class, then the question emerges whether this inclusion is proper. One reviewer of an earlier version of this paper gave the proper hint: While the graph languages generated by hyperedge replacement grammars have bounded tree width, the set of square grids has unbounded tree width and consequently the set of pseudotori, too.

Proposition 5. *The set of pseudotori cannot be generated by a hyperedge replacement grammar.*

Proof. Using the Propositions 4.7, 4.13 and 4.27 in Courcelle and Engelfriet [11], the language of a hyperedge replacement grammar turns out to have bounded tree width while the set of square grids has unbounded tree width as shown in Example 2.56. If one merges the corresponding leftmost and rightmost nodes as well as the upmost and the downmost ones, then one gets a set of tori still with unbounded tree width. As this is a subset of the set of pseudotori, the latter set has also unbounded tree width. This proves the statement.

7   Conclusion

In this paper, we have introduced fusion grammars as devices for the generation of hypergraph languages. The rules do nothing else than the fusion of nodes. Besides the nodes and hyperedges of the start hypergraph, one gets new nodes and hyperedges by copying connected components of intermediate derived hypergraphs, rather than by rule application as in most other grammar concepts. The first study shows that fusion grammars have some promising properties. For example, the membership problem is decidable and the well-known hyperedge replacement grammars can be simulated. Further research may prove the significance of the approach.

There are many possible ways to generalize fusion grammars. Fusion rules may be provided with positive and negative application conditions. The grammars may be provided with control conditions to cut down the totally free order of fusions and multiplications. The elements of the generated language may be filtered from the derived hypergraphs by other means than connectedness, terminality and removal of markers. It would be nice to overcome the limitation of generating connected components.

Clearly, to get a deeper insight into fusion grammars, one can start with the usual considerations of grammatical structures like closure properties, further decidability and undecidability properties and further relationships to other classes of grammars like, for example, context-sensitive hyperedge replacement grammars and monotone graph grammars. A further interesting candidate for comparison may be membrane systems introduced by Păun [12].

Fusion grammars allow massive parallelism where not only the derivation process may be massively parallel, but also many members of the generated language can be derived in parallel. Can one use this fact advantageously?

Acknowledgment. We are greatful to the anonymous reviewers for their valuable comments. To one of them we owe the idea to show Proposition 5.

References

1. Păun, G., Rozenberg, G., Salomaa, A.: DNA Computing - New Computing Paradigms. Springer, Heidelberg (1998)
2. Adleman, L.M.: Molecular computation of solutions to combinatorial problems. Science **266**, 1021–1024 (1994)
3. Peitgen, H.-O., Jürgens, H., Saupe, D.: Chaos and Fractals - New Frontiers of Science, 2nd edn. Springer, New York (2004)
4. Grünbaum, B., Shephard, G.C.: Tilings and Patterns. W. H. Freeman and Company, New York (1987)
5. Kreowski, H.-J., Kuske, S.: Graph multiset transformation - a new framework for massively parallel computation inspired by DNA computing. Nat. Comput. **10**(2), 961–986 (2011). doi:10.1007/s11047-010-9245-6
6. Andersen, J.L., Flamm, C., Merkle, D., Stadler, P.F.: A software package for chemically inspired graph transformation. In: Echahed, R., Minas, M. (eds.) ICGT 2016. LNCS, vol. 9761, pp. 73–88. Springer, Cham (2016). doi:10.1007/978-3-319-40530-8_5
7. Kreowski, H.-J., Klempien-Hinrichs, R., Kuske, S.: Some essentials of graph transformation. In: Esik, Z., Martin-Vide, C., Mitrana, V. (eds.) Recent Advances in Formal Languages and Applications. Studies in Computational Intelligence, vol. 25, pp. 229–254. Springer (2006)
8. Ehrig, H., Kreowski, H.-J.: Parallelism of manipulations in multidimensional information structures. In: Mazurkiewicz, A. (ed.) MFCS 1976. LNCS, vol. 45, pp. 284–293. Springer, Heidelberg (1976). doi:10.1007/3-540-07854-1_188
9. Drewes, F., Habel, A., Kreowski, H.-J.: Hyperedge replacement graph grammars. In: Rozenberg, G. (ed.) Handbook of Graph Grammars and Computing by Graph Transformation, vol. 1: Foundations, Chap. 2, pp. 95–162. World Scientific (1997)
10. Habel, A.: Hyperedge Replacement: Grammars and Languages. LNCS, vol. 643. Springer, Heidelberg (1992). doi:10.1007/BFb0013875
11. Courcelle, B., Engelfriet, J.: Graph Structure and Monadic Second-Order Logic - A Language-Theoretic Approach. Encyclopedia of Mathematics and Its Applications, vol. 138. Cambridge University Press, Cambridge (2012)
12. Păun, G.: Computing with membranes. J. Comput. Syst. Sci. **61**(1), 108–143 (2000)

Predictive Shift-Reduce Parsing
for Hyperedge Replacement Grammars

Frank Drewes[1], Berthold Hoffmann[2], and Mark Minas[3(✉)]

[1] Umeå universitet, Umeå , Sweden
drewes@cs.umu.se
[2] Universität Bremen, Bremen, Germany
hof@informatik.uni-bremen.de
[3] Universität der Bundeswehr München, Neubiberg, Germany
mark.minas@unibw.de

Abstract. Graph languages defined by hyperedge replacement grammars can be NP-complete. We study predictive shift-reduce (PSR) parsing for a subclass of these grammars, which generalizes the concepts of SLR(1) string parsing to graphs. PSR parsers run in linear space and time. In comparison to the predictive top-down (PTD) parsers recently developed by the authors, PSR parsing is more efficient and more general, while the required grammar analysis is easier than for PTD parsing.

Keywords: Hyperedge replacement grammar · Graph parsing · Grammar analysis

1 Introduction

"It is well known that hyperedge replacement (HR, see [11]) can generate NP-complete graph languages [1]. In other words, even for fixed HR languages parsing is hard. Moreover, even if restrictions are employed that guarantee L to be in P, the degree of the polynomial usually depends on L; see [16].[1] Only under rather strong restrictions the problem is known to become solvable in cubic time [5,21]." This quote is from our paper [8] on predictive top-down (PTD) parsing, an extension of $SLL(1)$ string parsing [17] to HR graph grammars [11]. The parser generator has been extended to the contextual HR grammars devised in [6,7]; it approximates Parikh images of auxiliary grammars in order to determine whether a grammar is PTD-parsable [9], and generates parsers that run in quadratic time, and in many cases in linear time.

Here we devise—somewhat complementary—efficient bottom-up parsers for HR grammars, called *predictive shift-reduce (PSR) parsers*, which extend $SLR(1)$ parsers [4], a member of the $LR(k)$ family of deterministic bottom-up parsers for context-free grammars [15]. We describe how PSR parsers work and how they can be constructed, and relate them to $SLR(1)$ string and PTD graph parsers.

---

[1] This result has been exploited for parsing natural language in the system *Bolinas* [2].

© Springer International Publishing AG 2017
J. de Lara and D. Plump (Eds.): ICGT 2017, LNCS 10373, pp. 106–122, 2017.
DOI: 10.1007/978-3-319-61470-0_7

In Sect. 2 we recall basic notions of HR grammars. We sketch $SLR(1)$ string parsing in Sect. 3 and describe in Sect. 4 how it can be lifted to PSR parsing of graphs with HR grammars. Then, in Sect. 5, we describe how HR grammars can be analysed for being PSR-parsable. Section 6 is devoted to the discussion of related work. Further work is outlined in Sect. 7.

2   Hyperedge Replacement Grammars

We let $\mathbb{N}$ denote the non-negative integers. A^* denotes the set of all finite sequences over a set A; the empty sequence is denoted by ε, the length of a sequence α by $|\alpha|$. For a function $f\colon A \to B$, its extension $f^*\colon A^* \to B^*$ to sequences is defined by $f^*(a_1 \cdots a_n) = f(a_1) \cdots f(a_n)$, for all $a_1, \dots, a_n \in A$, $n \geqslant 0$.

We consider an alphabet Σ of symbols for labeling edges that comes with an *arity* function $arity\colon \Sigma \to \mathbb{N}$. The subset $\mathcal{N} \subseteq \Sigma$ is the set of *nonterminal labels*.

An *edge-labeled hypergraph* $G = \langle \dot{G}, \bar{G}, att_G, \ell_G \rangle$ over Σ (a *graph*, for short) consists of disjoint finite sets $\dot{G}$ of *nodes* and $\bar{G}$ of *hyperedges* (*edges*, for short) respectively, a function $att_G\colon \bar{G} \to \dot{G}^*$ that *attaches* sequences of nodes to edges, and a *labeling* function $\ell_G\colon \bar{G} \to \Sigma$ so that $|att_G(e)| = arity(\ell_G(e))$ for every edge $e \in \bar{G}$. Edges are said to be *nonterminal* if they carry a nonterminal label, and *terminal* otherwise; the set of all graphs over Σ is denoted by $\mathcal{G}_\Sigma$. A *handle graph* G for $A \in \mathcal{N}$ consists of just one edge x and $k = arity(A)$ pairwise distinct nodes $n_1, \dots, n_k$ such that $\ell_G(x) = A$ and $att_G(x) = n_1 \dots n_k$.

Given graphs G and H, a *morphism* $m\colon G \to H$ is a pair $m = \langle \dot{m}, \bar{m} \rangle$ of functions $\dot{m}\colon \dot{G} \to \dot{H}$ and $\bar{m}\colon \bar{G} \to \bar{H}$ that preserves labels and attachments: $\ell_H \circ \bar{m} = \ell_G$, and $att_H \circ \bar{m} = \dot{m}^* \circ att_G$ (where "$\circ$" denotes function composition). A morphism $m\colon G \to H$ is *injective* and *surjective* if both $\dot{m}$ and $\bar{m}$ have the respective property. If m is injective and surjective, it makes G and H *isomorphic*. We do not distinguish between isomorphic graphs.

Definition 1 (HR Rule). A *hyperedge replacement rule* (*rule*, for short) $r = L \to R$ consists of graphs L and R over Σ such that the *left-hand side* L is a handle graph with $\dot{L} \subseteq \dot{R}$.

Let r be a rule as above, and consider some graph G. An injective morphism $m\colon L \to G$ is called a *matching* for r in G. The *replacement* of $m(L)$ by R is then given as the graph H, which is obtained from the disjoint union of G and R by removing the single edge in $m(\bar{L})$ and identifying, for every node $v \in \dot{L}$, the nodes $m(v) \in \dot{G}$ and $v \in \dot{R}$. We then write $G \Rightarrow_{r,m} H$ (or just $G \Rightarrow_r H$) and say that H is *derived* from G by r.

The notion of rules introduced above gives rise to the class of HR grammars.

Definition 2 (HR Grammar [11]). A *hyperedge-replacement grammar* (*HR grammar*, for short) is a triple $\Gamma = \langle \Sigma, \mathcal{R}, Z \rangle$ consisting of a finite labeling alphabet Σ, a finite set $\mathcal{R}$ of rules, and a start graph $Z \in \mathcal{G}_\Sigma$.

We write $G \Rightarrow_\mathcal{R} H$ if $G \Rightarrow_{r,m} H$ for some rule $r \in \mathcal{R}$ and a matching $m\colon L \to G$, and denote the transitive-reflexive closure of $\Rightarrow_\mathcal{R}$ by $\Rightarrow_\mathcal{R}^*$. The language generated by Γ is given by $\mathcal{L}(\Gamma) = \{G \in \mathcal{G}_{\Sigma \setminus \mathcal{N}} \mid Z \Rightarrow_\mathcal{R}^* G\}$.

Without loss of generality, we assume that the start graph Z consists of a single edge labeled with a symbol $S \in \mathcal{N}$ of arity 0, that it is the left-hand side of just one rule, and that it does not occur in any right-hand side.

Graphs are drawn as in Example 1. Circles represent nodes, and boxes of different shapes represent edges. The box of an edge contains its label, and is connected to the circles of its attached nodes by lines; these lines are ordered clockwise around the edge, starting to its left. Terminal edges with two attached nodes are usually drawn as arrows from their first to their second attached node, and the edge label is ascribed to that arrow (but omitted if there is just one label, as in Example 1 below). In rules, identifiers like "x" at nodes identify corresponding nodes on the left-hand and right-hand sides.

Example 1. With a start graph as assumed above, the HR grammar below derives n-ary trees, like the graph on the right:

3   Shift-Reduce Parsing of Strings

Our shift-reduce parser for HR grammars borrows and extends concepts known from the family of context-free $LR(k)$ parsers [15], which is why we recall these concepts first. As context-free grammars, shift-reduce parsing, and in particular $LR(k)$ parsing appear in every textbook on compiler construction, we discuss these matters just at hand of an example.

Example 2. The *Dyck language* of matching nested parentheses consists of strings over the symbols "[" and "]"; it can be defined by a context-free string grammar with four rules $\mathcal{D} = \{S \rightarrow T, T \rightarrow [B], B \rightarrow TB, B \rightarrow \varepsilon\}$, to which we refer by the numbers 0 to 3; S, T, and B are nonterminals, and ε denotes the empty string.

Starting with the string consisting only of S, the rules are applied to strings of nonterminals and terminals, by replacing an occurrence of their left-hand side by their right-hand side; this done repeatedly until all nonterminals have been replaced. So we can derive a word of the Dyck language:

$$S \underset{0}{\Rightarrow} T \underset{1}{\Rightarrow} [B] \underset{2}{\Rightarrow} [TB] \underset{3}{\Rightarrow} [T] \underset{1}{\Rightarrow} [[B]] \underset{3}{\Rightarrow} [[\,]] \tag{1}$$

A context-free *parser* checks whether a string like "[[]]" belongs to the language of a grammar, and constructs a derivation as above if this is the case. A parser is modeled by a stack automaton that reads an input string from left to right, and uses a stack for remembering its actions. In a (general) shift-reduce parser, a configuration can be represented as $\alpha \cdot w$, where w is the unconsumed input, a terminal string, and α is the stack, consisting of the nonterminal and terminal symbols that have been parsed so far. (The rightmost symbol is the "top".) The parser has its name from the two kinds of actions it performs (where α is an arbitrary string of symbols, and w an arbitrary terminal string):

- *Shift* consumes the first terminal symbol a of the input, and pushes it onto the stack. Our parser does shifts for parentheses:

$$\alpha \boldsymbol{.} [w \vdash \alpha[\boldsymbol{.} w \qquad \alpha \boldsymbol{.}]w \vdash \alpha]\boldsymbol{.} w$$

- *Reduce* pops the right-hand side symbols of a production from the stack, and pushes its left-hand side onto it. Our parser has the reductions (for symbol sequences α and terminal symbol sequences w):

$$T\boldsymbol{.} \vdash_0 S\boldsymbol{.} \qquad \alpha[B]\boldsymbol{.} w \vdash_1 \alpha T\boldsymbol{.} w \qquad \alpha TB\boldsymbol{.} w \vdash_2 \alpha B\boldsymbol{.} w \qquad \alpha\boldsymbol{.} w \vdash_3 \alpha B\boldsymbol{.} w$$

If the rule is the start rule, and α and w are empty, the parser terminates, and *accepts* the word, as in the first reduction.

A successful *parse* of a string w is a sequence of shift and reduce actions starting from the *initial configuration* $\varepsilon\boldsymbol{.} w$ to the *accepting configuration* $S\boldsymbol{.}\varepsilon$, as below:

$$\boldsymbol{.}[[]] \vdash [\boldsymbol{.}[]] \vdash [[\boldsymbol{.}]] \vdash_3 [[B\boldsymbol{.}]] \vdash [[B]\boldsymbol{.}] \vdash_1 [T\boldsymbol{.}] \vdash_3 [TB\boldsymbol{.}] \vdash_2 [B\boldsymbol{.}] \vdash [B]\boldsymbol{.} \vdash_1 T\boldsymbol{.} \vdash_0 S\boldsymbol{.}$$

The reduction steps of a successful parse, read in reverse, yield a rightmost derivation of "[[]]" from S, in this case the one in (1) above.

This parser is *nondeterministic*: E.g., in the configuration "$[TB\boldsymbol{.}]$", the following actions are possible:

1. a reduction with rule $B \to TB$, leading to the configuration $[B\boldsymbol{.}]$;
2. a reduction with rule $B \to \varepsilon$, leading to the configuration $[TBB\boldsymbol{.}]$; or
3. a shift of the symbol "]", leading to the configuration $[TB]\boldsymbol{.}$.

Only action 1 will lead to the successful parse above. After action 2, further reduction is impossible, even after a subsequent shift of the unconsumed "]"; after action 3, no further action is possible. In such situations, the parser must *backtrack*, i.e., undo actions and try alternative ones, until it can accept the word, or fails altogether.

Since backtracking is inefficient, shift-reduce parsers are extended by two concepts so that they can predict which action in a configuration will lead to a successful parse:

- A *lookahead* of $k > 0$ input symbols helps to decide for an action. In the situation sketched above, the reductions 1 and 2 should only be done if the first input symbol is "]", which is the only terminal symbol that may follow B in the derivations with the grammar.

- A *characteristic finite automaton* (CFA) controls the order in which actions are performed; in the configuration $\alpha[T\,B\,\boldsymbol{.}]$, the CFA should indicate that rule $B \to T\,B$ shall be reduced, not rule $B \to \varepsilon$.

Different lengths of lookahead, and several notions of CFAs can be used to construct a predictive shift-reduce parser. The most general one is Knuth's $LR(k)$ method [15]; here we just consider the simplest case of DeRemer's $SLR(k)$ parser [4], namely for a single symbol of lookahead, i.e., $k = 1$.

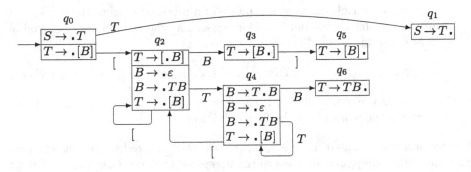

Fig. 1. $SLR(1)$ automaton $A_{\mathcal{D}}$ of the Dyck grammar

The transition diagram of the CFA $A_{\mathcal{D}}$ for the Dyck language is shown in Fig. 1. It is constructed as follows. The nodes q_0 to q_6 define its *states*, which are characterized by sets of so-called items; an *item* is a rule with a dot between the symbols on the right-hand side; e.g., state q_3 is characterized by the single item $T \to [B\,\boldsymbol{.}]$; in this state, the parser has recognized the symbols [and B of rule $T \to [B]$, but not the closing parenthesis. The transitions $q \xrightarrow{x} q'$ define successor state q' of a state q after recognizing a symbol x.

The start state q_0 is described by the item $S \to \boldsymbol{.}T$, which is called a *kernel item*. Since recognizing the nonterminal T implies to recognize the rule of T, $T \to \boldsymbol{.}[B]$ is the *closure item* of this state. The symbols appearing right of the dot in state q_0 can be recognized next; so, state q_0 has two transitions: under the nonterminal T to state q_1 with the kernel item $S \to T\,\boldsymbol{.}$ (T is being read), and under the terminal "[" to state q_2 with the kernel item $T \to [\boldsymbol{.}B]$. State q_2 has closure items $B \to \boldsymbol{.}\varepsilon$ and $B \to \boldsymbol{.}TB$, and the latter item has a further closure item $T \to \boldsymbol{.}[B]$. While state q_1 has no transitions (nothing more needs to be recognized), state q_2 has three successor states, under the nonterminals B, T, and the terminal "[". The transition under "[" loops on state q_2. The remaining states and transitions are determined analogously.

The stack of the $SLR(1)$ parser is extended to contain an alternating sequence of states and symbols, e.g., "$q_0[q_2[q_2Tq_4Tq_4$", which record a path $q_0 \xrightarrow{[} q_2 \xrightarrow{[}$ $q_2 \xrightarrow{T} q_4 \xrightarrow{T} q_4$ in its CFA $A_{\mathcal{D}}$, starting in the initial state. The actions of the parser are determined by its actual (topmost) state, and are modified wrt. those of the nondeterministic parser as follows:

- *Shift* consumes the first terminal a of the input, and pushes it onto the stack, with the successor state q' so that $q \overset{a}{\to} q'$ is in $A_\mathcal{D}$. For our grammar, and $i \in \{0, 2, 4\}$:

$$\alpha q_3 \bullet]w \vdash \alpha q_3]q_5 \bullet w \qquad \alpha q_i \bullet [w \vdash \alpha q_i[q_2 \bullet w$$

- *Reduce* pops the right-hand side of a production $A \to \beta$ (and the intermediate states) from the stack, but only if the lookahead—the first input symbol—may follow A in derivations, and pushes the left-hand side A, and the successor state q' so that $q \overset{A}{\to} q'$. If $A = S$ and the input is empty, the parser accepts the word. For our grammar:

$$q_0 T q_1 \bullet \vdash_0 S \qquad\qquad \alpha q_0[q_2 B q_3]q_5 \bullet w \vdash_1 \alpha q_0 T q_1 \bullet w$$

$$\alpha q_2[q_2 B q_3]q_5 \bullet w \vdash_1 \alpha q_2 T q_4 \bullet w \qquad \alpha q_4[q_2 B q_3]q_5 \bullet w \vdash_1 \alpha q_4 T q_4 \bullet w$$

$$\alpha q_2 T q_4 B q_6 \bullet]w \vdash_2 \alpha q_2 B q_3 \bullet]w \qquad \alpha q_4 T q_4 B q_6 \bullet]w \vdash_2 \alpha q_4 B q_6 \bullet]w$$

$$\alpha q_2 \bullet]w \vdash_3 \alpha q_2 B q_3 \bullet]w \qquad\qquad \alpha q_4 \bullet]w \vdash_3 \alpha q_4 B q_6 \bullet]w$$

The CFA may reveal conflicts for predictive parsing:

- If a state allows to shift some terminal a, and a reduction under the lookahead a, this is a *shift-reduce conflict*.
- If a state allows reductions of different productions under the same lookahead symbol, this is a *reduce-reduce conflict*.

Whenever the automaton is conflict-free, the $SLR(1)$ parser exists, and is deterministic. The automaton $A_\mathcal{D}$ for the Dyck language is conflict-free: In states q_2 and q_4, rule $B \to \varepsilon$ can be reduced if the input begins with the only follower symbol "]" of B, which is not in conflict with the shift transitions from these states under "[".

A deterministic parse with the $SLR(1)$ parser for $\mathcal{D}$ is as follows:

$$q_0 \bullet [[]] \vdash q_0[q_2 \bullet []] \vdash q_0[q_2[q_2 \bullet]] \underset{3}{\vdash} q_0[q_2[q_2 B q_3 \bullet]] \vdash q_0[q_2[q_2 B q_3]q_5 \bullet]$$

$$\underset{1}{\vdash} q_0[q_2 T q_4 \bullet] \underset{3}{\vdash} q_0[q_2 T q_4 B q_6 \bullet] \underset{2}{\vdash} q_0[q_2 B q_3 \bullet] \vdash q_0[q_2 B q_3]q_5 \bullet$$

$$\underset{1}{\vdash} q_0 T q_1 \bullet \qquad \underset{0}{\vdash} S \text{ (accept)}$$

4   Predictive Shift-Reduce Parsing for HR Grammars

We shall now transfer the basic ideas of shift-reduce parsing to HR grammars. First we define a textual notation for graphs and HR rules. A graph G can be represented as a pair $u_G = \langle s, \dot{G} \rangle$, called *(graph) clause* of G, where s is a sequence of *(edge) literals* $a(x_1, \ldots, x_k)$, one for every edge $e \in \bar{G}$ with $\ell_G(e) = a$ and $att_G(e) = x_1 \ldots x_k$. When writing down u_G, we omit $\dot{G}$ and write just s if $\dot{G}$ is clear from the context. For an HR rule $L \to R$, the *rule clause* is $u_L \to u_R$, with $\dot{L} \subseteq \dot{R}$.

While the order of literals in a graph clause is irrelevant, we shall process the literals on the right-hand side of a rule clause in the given order.[2]

Example 3 (Tree Rule Clauses and Tree Clauses). The rules of the tree grammar in Example 1 are represented by the clauses

$$S() \to T(x) \qquad T(y) \to T(y)\, edge(y, z)\, T(z) \qquad T(y) \to \varepsilon$$

We shall refer to them by r_1, r_2, r_3. The empty sequence ε in the last rule is a short-hand for the clause $\langle \varepsilon, \{y\} \rangle$. One of the possible clauses representing the graph in Example 1 is "$edge(1, 2)\, edge(1, 3)\, edge(2, 4)\, edge(2, 5)$".

We will use this simple example to demonstrate the basic ideas of PSR parsing.

The PSR graph parser will use configurations $\alpha . w$, and rely on a CFA for its control, just like an $SLR(1)$ parser. However, instead of just symbols, the configurations will consist of literals, and something has to be done in order to properly determine the nodes of these literals in the host graph. This makes the construction more complicated.

If we disregard for a moment the assignment of host graph nodes to the literals, the states of the CFA are defined as sets of items, i.e., of rule clauses with a dot at some place in their right-hand side. Consider the kernel item $T(y) \to T(y)\, edge(y, z) . T(z)$ as an example. It has closure items of the form $T(y) \to .T(y) edge(y, z)\, T(z)$ and $T(y) \to .$. However, we have to take care of the node names: Since the closure is built according to the literal $T(z)$, the y in the closure items is actually the z of the kernel item, and their z is a node not in the kernel item at all. This has to be reflected in the closure items, without causing name clashes. Our method will be the following: First we distinguish those nodes in the kernel items to which nodes of the host graph will have been assigned when the state is entered. These are called the *parameters* of the state. In the present state – let us call it q_2 – the parameters will correspond to y and z since the literals to the left of the dot are already on the stack in this state. First we replace y and z by parameter names, say $\boldsymbol{x}$ and $\boldsymbol{y}$, in the kernel item. Then we rename the nodes on the left-hand side of the closure items according to the kernel literal causing the closure, i.e., we replace y by $\boldsymbol{y}$ in the closure items. The remaining node names are preserved – they correspond to nodes that have not yet been assigned any host graph nodes and are thus not parameters.[3] We now call this state $q_2(\boldsymbol{x}, \boldsymbol{y})$ to indicate that $\boldsymbol{x}$ and $\boldsymbol{y}$ are its formal parameters which have to be instantiated by concrete host graph nodes whenever the parser enters the state.

---

[2] We assume that this order is provided with the HR grammar. Finding an appropriate order for PSR parsing automatically can be done by dataflow analysis, but is outside the scope of this paper.

[3] In general, we may have to introduce fresh names for non-parameter nodes in the closure items as well in order to avoid name clashes, but this is not necessary in the present example.

State $q_2(x, y)$ now gets a transition under the literal $T(y)$ to a state, let us call it $q_3(x, y)$, which has two kernel items

$$T(x) \to T(x) \, edge(x, y) \, T(y) \, . \text{ and } T(y) \to T(y) \, . \, edge(y, z) \, T(z).$$

This state also gets x and y as formal parameters. For the transition, we specify which nodes of $q_2(x, y)$ define the parameters in $q_3(x, y)$, writing it in the form of a "call" $q_3(x, y)$. (Note that x and y are the parameters in state $q_2(x, y)$ that are thus transferred to the (equally named) formal parameters of $q_3(x, y)$.)

The second item of $q_3(x, y)$ causes a transition under $edge(y, z)$ to a state that would have a kernel item $T(y) \to T(y) edge(y, z) . T(z)$, where the actual parameter z is determined by the shift. However, this kernel item equals the one of $q_2(x, y)$ up to the names of formal parameters so that we identify these states, and "call" $q_2(x, y)$, specifying its actual parameters by writing $q_2(y, z)$ on the transition. In general, a transition is thus defined by a literal, and by a call that defines the actual parameters, thereby passing nodes from one state to the other.

Figure 2 shows the CFA for the tree grammar, built according to these considerations. (Note that $q_2(x, y)$ and $q_3(x, y)$ are as discussed above.) A special case arises in the start state $q_0(x)$. In order to work without backtracking, parsing has to start at nodes of the start rule that can be uniquely determined before parsing starts. In our example, the node u of the host graph that corresponds to x in the start item $S() \to . T(x)$ must be determined, and assigned to the formal parameter x of $q_0(x)$ before running the parser. This is done by a procedure devised in [9, Sect. 4], computing the possible incidences of all nodes created by a grammar; only if the start node u can be distinguished from all other nodes

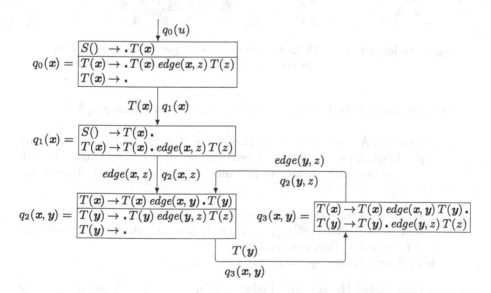

Fig. 2. The PSR CFA for the tree grammar Example 3

generated by the grammar, predictive parsing is possible. In our example, u is the unique node in the input graph that has no incoming edges, i.e., the root of the tree. (If the input graph has more than one root, or no root at all, it cannot be a tree, and parsing fails immediately.) So the start item is renamed to $S() \to .T(\boldsymbol{x})$, and $q_0(\boldsymbol{x})$ is entered with the call $q_0(u)$.

Figure 4 shows moves of a PSR parser that accept the tree of Fig. 3 in state q_1^1 that indicates a reduction with the start rule. We are using a compact form to denote concrete instances of states (i.e., with actual parameters being assigned to them): for a state $q(\boldsymbol{x}_1, \ldots, \boldsymbol{x}_k)$ and an assignment $\mu: \{\boldsymbol{x}_1, \ldots, \boldsymbol{x}_k\} \to G$ we let q^μ denote $q(\mu(\boldsymbol{x}_1), \ldots, \mu(\boldsymbol{x}_k))$. Moreover, in Fig. 4 we just denote μ by a list of nodes, i.e., q_0^1 denotes q_0^μ where $\mu(\boldsymbol{x}) = 1$. We use a similar shorthand for literals, e.g., e^{12} and T^1 abbreviate literals $edge(1,2)$ and $T(1)$, respectively.

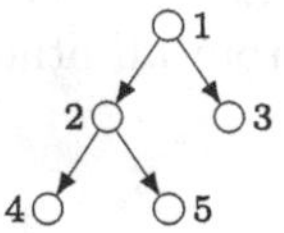

$$
\begin{aligned}
& q_0^1 \cdot e^{12} e^{13} e^{24} e^{25} \\
\vdash_3 \quad & q_0^1 T^1 \underline{q_1^1} \cdot e^{12} e^{13} e^{24} e^{25} \\
\vdash \quad & q_0^1 T^1 q_1^1 e^{12} q_2^{12} \cdot e^{13} e^{24} e^{25} \\
\vdash_3 \quad & q_0^1 T^1 q_1^1 e^{12} q_2^{12} T^2 \underline{q_3^{12}} \cdot e^{13} e^{24} e^{25} \\
\vdash \quad & q_0^1 T^1 q_1^1 e^{12} q_2^{12} T^2 q_3^{12} e^{24} q_2^{24} \cdot e^{13} e^{25} \\
\vdash_3 \quad & q_0^1 T^1 q_1^1 e^{12} q_2^{12} T^2 q_3^{12} e^{24} q_2^{24} T^4 \underline{q_3^{24}} \cdot e^{13} e^{25} \\
\vdash_2 \quad & q_0^1 T^1 q_1^1 e^{12} q_2^{12} T^2 q_3^{12} \cdot e^{13} e^{25} \\
\vdash \quad & q_0^1 T^1 q_1^1 e^{12} q_2^{12} T^2 q_3^{12} e^{25} q_2^{25} \cdot e^{13} \\
\vdash_3 \quad & q_0^1 T^1 q_1^1 e^{12} q_2^{12} T^2 q_3^{12} e^{25} q_2^{25} T^5 \underline{q_3^{25}} \cdot e^{13} \\
\vdash_2 \quad & q_0^1 T^1 q_1^1 e^{12} q_2^{12} T^2 q_3^{12} \cdot e^{13} \\
\vdash_2 \quad & q_0^1 T^1 q_1^1 \cdot e^{13} \\
\vdash \quad & q_0^1 T^1 q_1^1 e^{13} q_2^{13} \cdot \\
\vdash_3 \quad & q_0^1 T^1 q_1^1 e^{13} q_2^{13} T^3 \underline{q_3^{13}} \cdot \\
\vdash_2 \quad & q_0^1 T^1 q_1^1 \cdot
\end{aligned}
$$

Fig. 3. An input tree

Fig. 4. Moves of a PSR parser recognizing Fig. 3; places where reductions occur are underlined

The operations of the PSR parser work as follows on a host graph G:

Shift. Let the CFA contain a transition from state $p(\boldsymbol{x}_1, \ldots, \boldsymbol{x}_k)$ to state $q(\boldsymbol{y}_1, \ldots, \boldsymbol{y}_l)$ labeled by the terminal edge literal $e(v_1, \ldots, v_m)$ and the call $q(u_1, \ldots, u_l)$, and consider a concrete instance p^μ of $p(\boldsymbol{x}_1, \ldots, \boldsymbol{x}_k)$. Then there is a *shift* from p^μ to q^ν if

1. μ can be extended to the non-parameter nodes among $v_1, \ldots, v_m$, yielding an assignment μ' such that $e(\mu'(v_1), \ldots, \mu'(v_m))$ is a hitherto unconsumed edge literal of G (which is thus consumed) and
2. ν is defined by setting $\nu(\boldsymbol{y}_i) = \mu'(u_i)$ for $i = 1, \ldots, l$.

The shift then pushes the consumed edge $e(\mu'(v_1), \ldots, \mu'(v_m))$ and q^ν onto the stack.

Reduce. Let the topmost stack element be the state $q_k^{\nu_k}$, and assume that it contains an item of the form $A(w) \to s_1(w_1) \cdots s_k(w_k)\,\bullet\,$. Thus, $w, w_1, \ldots, w_k$ are sequences of formal parameters of q_k and the stack is of the form

$$\cdots p^\mu s_1(w_1')\, q_1^{\nu_1} \cdots s_k(w_k')q_k^{\nu_k}$$

where $w_i' = \nu_k^*(w_i)$ for all i.

The CFA would then allow a transition from p^μ to consume the instantiated left-hand side $A(\nu_k^*(w))$. Let q^ν be the concrete target state of this transition. Then a *reduction* can be performed by popping $s_1(w_1')q_1^{\nu_1} \cdots s_k(w_k')q_k^{\nu_k}$ from the stack and pushing $A(\nu_k^*(w))$ and q^ν onto it.

Note that the parser has to choose the next action in states like q_1^ν or q_3^ν, which allow for a shift, but also for a reduction with rule r_1 and r_2, respectively. The parser predicts the next step by inspecting the unconsumed edges. We will discuss this in the next section. Note also that the PSR shift step differs from its $SLR(1)$ counterpart in an important detail: while the string parser just reads the uniquely determined next symbol of the input word, the graph parser must choose an appropriate edge. There may be states with several outgoing shift transitions (which do not occur in the present example), and the PSR parser has to choose between them. The parser, when it has selected a specific shift transition, may even have to choose between several edges to shift. E.g., in Fig. 4 the second step could shift the edge $edge(1,3)$ instead of $edge(1,2)$. We will discuss this problem below when considering shift-shift as well as other conflicts and the *free edge choice* property.

5   Predictive Shift-Reduce Parsability

A CFA can be constructed for every HR grammar; the general procedure works essentially as described above. However, one must usually deal with items whose left-hand sides still contain nodes which have not yet been located in the host graph. (Such items do not occur in the tree example.) We ignore this issue here due to space restrictions and refer to [8]. We shall now outline the three criteria for an HR grammar to be PSR-parsable (or just PSR for short):

1. *Neighbor-determined start nodes:* Prior to actual parsing, one must be able to determine the nodes where parsing will start. This has already been examined in [8,9] and is not considered in this paper.
2. *Conflict-freeness:* An $SLR(1)$ parser can predict the next action iff its CFA is conflict-free (cf. Sect. 3). We define a similar concept for PSR parsing in the following.
3. *Free edge choice:* The parser, when it has selected a specific shift transition, may have to choose between several edges matching the edge pattern, as described above. A grammar has the *free edge choice* property if the parser can always choose freely between these edges. There are sufficient conditions for this property that can be effectively tested when testing for conflict-freeness. This has already been examined in [8]; so we do not discuss it here.

We shall now define conflict-freeness in PSR parsing similar to $SLR(1)$ parsing so that conflict-freeness implies that the PSR parser can always predict the next action. A graph parser, different from a string parser, must choose the next edge to be consumed from a set of appropriate unconsumed edges. It depends on the next action of the parser which edges are appropriate. We define a conflict as the situation that an unconsumed edge is appropriate for an action, but could be consumed also if another action was chosen. Conflict-freeness just means that there are no conflicts. Obviously, conflict-freeness then allows to always predict the correct action.

We now discuss how to identify host edges that are appropriate for the action caused by an item. For this purpose, let us first define items in PSR parsing more formally: An item $I = \langle L \rightarrow \alpha \cdot \beta \mid P \rangle$ consists of an HR rule $L \rightarrow \alpha\beta$ in clause representation with a dot indicating a position in the right-hand side and the set P of parameters, i.e., those nodes in the item to which we have already assigned nodes in the host graph. These host nodes are not yet known when constructing the CFA and the PSR parser, but we can interpret parameters as abstract host nodes. A "real" host node assigned to a parameter during parsing is mapped to the corresponding abstract node. All other host nodes are mapped to a special abstract node $-$. Edges of the host graph are mapped to abstract edges being attached to abstract nodes, i.e., $P \cup \{-\}$, and each abstract edge can be represented by an *abstract (edge) literal* in the usual way. Note that the number of different abstract literals is finite because $P \cup \{-\}$ is finite.

Consider any valid host graph in $\mathcal{L}(\Gamma)$, represented by clause γ derived by the derivation $S = \alpha_1 \Rightarrow \cdots \Rightarrow \alpha_n = \gamma$. We assume that the ordering of edge literals is preserved in each derivation step. We then select any mapping of nodes in γ to abstract nodes $P \cup \{-\}$ such that no node in P is the image of two different host nodes. Edge literals are mapped to the corresponding abstract literals. The resulting sequence of literals can then be viewed as a derivation in a context-free string grammar $\Gamma(P)$ that can be effectively constructed from Γ in the same way as described in [9, Sect. 4]; details are omitted here because of space restrictions. $\Gamma(P)$ has the nice property that we can use this context-free string grammar instead of Γ to inspect conflicts. This is shown in the following.

Consider an item $I = \langle L \rightarrow \alpha \cdot \beta \mid P \rangle$. Each edge literal $e = l(n_1, \ldots, n_k)$ has the corresponding abstract literal $abstr_P(e) = l(m_1, \ldots, m_k)$ where $m_i = n_i$ if $n_i \in P$, and $m_i = -$ otherwise, for $1 \leqslant i \leqslant k$. Let us now determine all host edges, represented by their abstract literals, which can be consumed next if the action caused by this item is selected. The host edge consumed next must have the abstract literal $First_P(\beta) := abstr_P(e)$ if I is a shift item, i.e., β starts with a terminal literal e. If I, however, causes a reduction, i.e., $\beta = \varepsilon$, we can make use of $\Gamma(P)$. Any host edge consumed next must correspond to an abstract literal that is a follower of the abstract literal of L in $\Gamma(P)$. This is exactly the same concept as it is used for $SLR(1)$ parsing and indicated in Sect. 3. Let us use the notion $Follow_P(L)$ for this set of followers.

As an example, consider state $q_3(\boldsymbol{x}, \boldsymbol{y})$ in Fig. 2 with its items $\langle L_i \to \alpha_i \cdot \beta_i \mid P_i \rangle$, $i = 1, 2$, with $P_1 = \{x, y\}$ and $P_2 = \{y\}$. For the first item, one can compute

$$Follow_{P_1}(L_1) = Follow_{P_1}(T(x)) = \{edge(x, -), edge(-, -), \varepsilon\},$$

i.e., the host edge consumed next must either be an edge between the host node assigned to x and a node different from the ones assigned to x or y, or it must be a completely unrelated edge wrt. to the host nodes assigned to x or y, or all edges of the host edge have been completely consumed, indicated by ε.

For the second item, one can compute

$$First_{P_2}(\beta_2) = abstr_{P_2}(edge(y, z)) = edge(y, -),$$

i.e., the host edge under which the shift step is taken is an edge between the host node assigned to y and a node different from the ones assigned to x or y.

As $First_{P_2}(\beta_2)$ is not in $Follow_{P_1}(L_1)$ the edge under which the shift step is taken cannot be consumed next when the reduce step is taken instead. This condition is sufficient to avoid a conflict in $SLR(1)$ parsing, but this is not the case for PSR parsing. Because host edges are not consumed in a fixed sequence, $edge(y, -)$ might be consumed later when the reduce step is taken. We must actually compute the set of all abstract literals that may follow $abstr_{P_1}(L)$ immediately, or later in $\Gamma(P)$. Let us denote the set of all such abstract literals $Follow_P^*(L)$, whose computation from $\Gamma(P)$ is straightforward. In this example, one can see that

$$Follow_{P_1}^*(L_1) = \{edge(x, -), edge(-, -), \varepsilon\}.$$

We conclude that, if the parser can find a host edge matching $edge(y, -)$, this edge can never be consumed if the reduce step is taken; the parser must select the shift step. If it cannot find such a host edge, it must select the reduce step.

Note that the test for a conflict (which we have not yet defined formally) is not symmetric: we have just checked whether $First_{P_2}(\beta_2) \notin Follow_{P_1}^*(L_1)$. But we could also check whether any abstract literal in $Follow_{P_1}(L_1)$, i.e., if the reduce step is taken, can be consumed later when the shift step is taken instead. Let us denote the set of these abstract literals as $Follow_{P_2}^*(L_2, \beta_2)$, which can be computed using $\Gamma(P)$ in a straightforward way. In the tree example, it is

$$Follow_{P_2}^*(L_2, \beta_2) = \{edge(y, -), edge(-, -)\},$$

i.e., $edge(-, -) \in Follow_{P_1}(L_1) \cap Follow_{P_2}^*(L_2, \beta_2)$. Thus the parser cannot predict the next step by just checking the existence of host edges matching abstract literals in $Follow_{P_1}(L_1)$. But this is insignificant, because it can predict the correct action based on the other test (in the "opposite direction").

Analogous arguments apply if two different shift actions or two different reduction actions are possible in a state. However, there are no such states in the tree example.

We are now ready to define conflicting items in PSR parsing. In order to simplify the definition, we refer to sets $Follow_P(I)$ and $Follow_P^*(I)$ for an item $I = \langle L \to \alpha \,.\, \beta \mid P \rangle$. If I is a shift item, we define

$$Follow_P(I) := \{First_P(\beta)\} \text{ and } Follow_P^*(I) := Follow_P^*(L, \beta).$$

If I is a reduce item, we define

$$Follow_P(I) := Follow_P(L) \text{ and } Follow_P^*(I) := Follow_P^*(L).$$

Definition 3 (Conflicting items). Let I_1 and I_2 be two items with sets P_1 and P_2 of parameters, respectively. I_1 and I_2 are in *conflict* iff

$$Follow_P(I_1) \cap Follow_P^*(I_2) \neq \varnothing \wedge Follow_P(I_2) \cap Follow_P^*(I_1) \neq \varnothing$$

where $P = P_1 \cap P_2$. The conflict is called a *shift-shift, shift-reduce,* or *reduce-reduce conflict* depending on the actions caused by I_1 and I_2.

The above considerations make clear that the parser can predict the next action correctly if all states of its CFA are conflict-free. They also make clear that the parser has to consider only a fixed number of abstract edge literals in any state to choose the next action, and the host edge to shift if a shift is chosen. However, each abstract literal may match several host edges. But proper preprocessing of the host graph allows to find an (arbitrary, because of the free edge choice property) unconsumed host edge in constant time. This preprocessing is linear in the size of the graph (in space and time). Because the number of actions of the parser is linear in the size of the input graph, it follows that PSR parsing is linear in the size of the host graph.

The *Grappa* tool implemented by the third author generates PSR parsers based on the construction of the PSR CFA and the analysis of the three criteria outlined above. Table 1 summarizes test results for some HR grammars. The columns under "Grammar" indicate the size of the grammar in terms of the maximal arity of nonterminals (A), number of nonterminals (N), number of terminals (T) and number of rules (R). The columns under "CFA" indicate the size of the generated CFA in terms of the number of states (S), the overall number of items (I) and the number of transitions (T). The number of conflicts in the CFA are shown in the columns below "Conflicts" that report of shift-shift (S/S), shift-reduce (S/R) and reduce-reduce conflicts (R/R). Note that the grammars without any conflicts are PSR, the others are not. The columns under "Analysis" report on the time in milliseconds needed for creating the CFA (CFA) and checking for conflicts (Confl.), on a MacBook Pro 2013 (2,7 GHz Intel Core i7, Java 1.8.0 and Scala 2.12.1).

Table 1. Test results of some HR grammars.

Example	Grammar				CFA			Conflicts			Analysis [ms]	
	A	N	T	R	S	I	T	S/S	S/R	R/R	CFA	Confl
Trees (Example 3)	1	2	1	3	4	10	4	–	–	–	93	38
$a^n b^n c^n$ [8]	4	3	3	5	14	22	14	–	–	–	130	31
Nassi-Shneiderman diagrams [19]	4	3	3	6	12	78	59	–	–	–	233	76
Palindromes (1)	2	2	2	7	12	32	19	–	–	–	142	36
Arithmetic expressions	2	4	5	7	12	34	22	–	–	–	137	49
Series-parallel graphs	2	2	1	4	7	63	32	12	4	–	179	61
Structured flowcharts	2	3	4	6	14	75	50	–	4	–	212	81

6   Comparison with Related Work

PSR parsing can be compared with $SLR(1)$ string parsing if we define the representation of strings as graphs, and of context-free grammars as HR grammars.

The *string graph* $w^\bullet$ of a string $w = a_1 \cdots a_n \in A^*$ (with $n \geqslant 0$) consists (in clausal form) of n edge literals $a_i(x_{i-1}, x_i)$ over $n+1$ distinct nodes $x_0, \ldots, x_n$. (The empty string ε is represented by an isolated node.) The HR rule of a context-free rule $A \to \alpha$ (where $A \in \mathcal{N}$ and $\alpha \in \Sigma^*$) is $A^\bullet \to \alpha^\bullet$. For the purpose of this section, we represent an ε-rule $A \to \varepsilon$ by a rule that maps both nodes of $A^\bullet$ to the only node in $\varepsilon^\bullet$. Such rules are called "*merging*" in [8]. Context-free grammars and HR grammars can be *cleaned*, i.e., transformed into equivalent grammars without ε-rules and merging rules, respectively; however, they may loose their $SLL(1)$ and PTD property, respectively.

The *string graph grammar* of a context-free grammar G with a finite set $\mathcal{P}$ of rules and a start symbol S is the HR grammar $G^\bullet = (\Sigma, \mathcal{P}^\bullet, S^\bullet)$ with the rules $\mathcal{P}^\bullet = \{A^\bullet \to \alpha^\bullet \mid A \to \alpha \in \mathcal{P}\}$. It is easy to see that the HR language of $G^\bullet$ is $\mathcal{L}(G^\bullet) = \{w^\bullet \mid w \in \mathcal{L}(G)\}$.

The following can easily be shown by inspection of the automata of string and HR grammars.

Proposition 1. *For an* $SLR(1)$ *grammar without* ε-rules, *the string graph grammar is PSR.*

This allows to establish the expected relation between PTD and PSR string graph grammars.

Theorem 1. *The clean version of a PTD string graph grammar is PSR.*

Proof. The main result of [12] states that the ε-cleaned version $\tilde{G}$ of an $SLL(1)$ grammar G is $SLR(1)$. This result can be lifted to string graph grammars as follows: By [8, Theorem 2], the string graph grammar $G^\bullet$ is PTD since G is $SLL(1)$. The string graph grammar $\tilde{G}^\bullet$ is the clean version of $G^\bullet$. Since $\tilde{G}$ is $SLR(1)$, $\tilde{G}^\bullet$ is PSR by Proposition 1.   □

The inclusion of $SLR(1)$ grammars is proper, as is the inclusion of $SLL(1)$ grammars in PTD grammars.

Corollary 1. *There are context-free languages that cannot be generated with an $SLR(1)$ grammar, but have a PSR string graph grammar.*

Proof. The language of *palindromes* over $V = \{a, b\}$, i.e., all words which read the same backward as forward, can be generated by the unambiguous grammar with rules $S \to P$ and $P \to a \mid aa \mid aPa \mid b \mid bb \mid bPb$. Since the language cannot be recognized by a deterministic stack automaton [20, Prop. 5.10], this language neither has an $LL(k)$ parser, nor an $LR(k)$ parser. However, the grammar is PTD by [8, Theorem 2], and ε-free so that it is PSR by Theorem 1.   □

For graph languages beyond string graphs, a comparison of PTD and PSR appears to be difficult: On the one hand, the tree grammar in Example 3 is left-recursive, and not PTD. (However, moving the *edge*-literal to the front in rule r_2 makes the grammar PTD.) On the other hand, PTD grammars with merging rules are not PSR, and it will be rather difficult to lift the main result of [12] to the general case of graph languages.

For early related work on efficient parsing algorithms, we quote from the conclusions of [8]: "Related work on parsing includes precedence graph grammars based on node replacement [10,14]. These parsers are linear, but fail for some PTD-parsable languages, e.g. the trees in Example 1. According to our knowledge, early attempts to implement LR-like graph parsers [18] have never been completed. Positional grammars [3] are used to specify visual languages, but can also describe certain HR grammars. They can be parsed in an LR-like fashion, but many decisions are deferred until the parser is actually executed. The CYK-style parsers for unrestricted HR grammars (plus edge-embedding rules) implemented in DiaGen [19] work for practical languages, although their worst-case complexity is exponential."

7   Conclusions

We have devised a predictive shift-reduce (PSR) parsing algorithm for HR grammars, along the lines of $SLR(1)$ string parsing. For string graphs, PSR is strictly more powerful than $SLR(1)$ and predictive top-down (PTD) parsing [8]. Checking PSR-parsability is complicated enough, but easier than for PTD, as we do not need to consider HR rules that merge nodes of their left-hand sides. PSR parsers also work more efficiently than PTD parsers, namely in linear vs. quadratic time. The reader is encouraged to download the *Grappa* generator of PTD and PSR parsers and to conduct own experiments.[4]

Like PTD, PSR parsing can be lifted to contextual HR grammars [6,7], a class of graph grammars that is more relevant for the practical definition of graph

---

[4] The *Grappa* tool is available at www.unibw.de/inf2/grappa; the examples mentioned in Table 1 can be found there as well.

languages. This remains as part of future work. We will also study whether the specification of priorities for rules, as in the yacc parser generator [13], can be lifted to PSR parsing. However, an extension of PSR corresponding to $SLR(k)$ or $LR(k)$ may be computationally too difficult. A still open challenge is to find a HR (or contextual HR) language that has a PSR parser, but no PTD parser. The corresponding example for $LL(k)$ and $LR(k)$ string languages exploits that strings are always parsed from left to right—the palindrome example shows that this is not the case for PTD and PSR parsers.

References

1. Aalbersberg, I., Ehrenfeucht, A., Rozenberg, G.: On the membership problem for regular DNLC grammars. Discrete Appl. Math. **13**, 79–85 (1986)
2. Chiang, D., Andreas, J., Bauer, D., Hermann, K.M., Jones, B., Knight, K.: Parsing graphs with hyperedge replacement grammars. In: Proceedings of the 51st Annual Meeting of the Association for Computational Linguistic. Long Papers, vol. 1, pp. 924–932 (2013)
3. Costagliola, G., De Lucia, A., Orefice, S., Tortora, G.: A parsing methodology for the implementation of visual systems. IEEE Trans. Softw. Eng. **23**, 777–799 (1997)
4. DeRemer, F.L.: Simple LR(k) grammars. Commun. ACM **14**(7), 453–460 (1971)
5. Drewes, F.: Recognising k-connected hypergraphs in cubic time. Theor. Comput. Sci. **109**, 83–122 (1993)
6. Drewes, F., Hoffmann, B.: Contextual hyperedge replacement. Acta Informatica **52**, 497–524 (2015)
7. Drewes, F., Hoffmann, B., Minas, M.: Contextual hyperedge replacement. In: Schürr, A., Varró, D., Varró, G. (eds.) AGTIVE 2011. LNCS, vol. 7233, pp. 182–197. Springer, Heidelberg (2012). doi:10.1007/978-3-642-34176-2_16
8. Drewes, F., Hoffmann, B., Minas, M.: Predictive top-down parsing for hyperedge replacement grammars. In: Parisi-Presicce, F., Westfechtel, B. (eds.) ICGT 2015. LNCS, vol. 9151, pp. 19–34. Springer, Cham (2015). doi:10.1007/978-3-319-21145-9_2
9. Drewes, F., Hoffmann, B., Minas, M.: Approximating Parikh images for generating deterministic graph parsers. In: Milazzo, P., Varró, D., Wimmer, M. (eds.) STAF 2016. LNCS, vol. 9946, pp. 112–128. Springer, Cham (2016). doi:10.1007/978-3-319-50230-4_9
10. Franck, R.: A class of linearly parsable graph grammars. Acta Informatica **10**(2), 175–201 (1978)
11. Habel, A.: Hyperedge Replacement: Grammars and Languages. LNCS, vol. 643. Springer, Heidelberg (1992). doi:10.1007/BFb0013875
12. Hoffmann, B.: Cleaned SLL(1) grammars are SLR(1). Technical Report 17–1, Studiengang Informatik, Universität Bremen (2017). http://www.informatik.uni-bremen.de/~hof/papers/sllr.pdf
13. Johnson, S.C.: Yacc: Yet another compiler-compiler. Computer Science Technical Report 32, AT&T Bell Laboratories (1975)
14. Kaul, M.: Practical applications of precedence graph grammars. In: Ehrig, H., Nagl, M., Rozenberg, G., Rosenfeld, A. (eds.) Graph Grammars 1986. LNCS, vol. 291, pp. 326–342. Springer, Heidelberg (1987). doi:10.1007/3-540-18771-5_62
15. Knuth, D.E.: On the translation of languages from left to right. Inf. Control **8**(6), 607–639 (1965)

16. Lautemann, C.: The complexity of graph languages generated by hyperedge replacement. Acta Informatica **27**, 399–421 (1990)
17. Lewis II, P.M., Stearns, R.E.: Syntax-directed transduction. J. ACM **15**(3), 465–488 (1968)
18. Ludwigs, H.J.: A LR-like analyzer algorithm for graphs. In: Wilhelm, R. (ed.) GI - 10. Jahrestagung, Proceedings of the Saarbrücken, 30 September - 2 Oktober 1980. Informatik-Fachberichte, vol. 33, pp. 321–335 (1980)
19. Minas, M.: Diagram editing with hypergraph parser support. In: Proceedings of the 1997 IEEE Symposium on Visual Languages (VL 1997), Capri, Italy, pp. 226–233 (1997)
20. Sippu, S., Soisalon-Soininen, E.: Parsing Theroy I: Languages and Parsing, EATCS Monographs in Theoretical Computer Science, vol. 15 (1988)
21. Vogler, W.: Recognizing edge replacement graph languages in cubic time. In: Ehrig, H., Kreowski, H.-J., Rozenberg, G. (eds.) Graph Grammars 1990. LNCS, vol. 532, pp. 676–687. Springer, Heidelberg (1991). doi:10.1007/BFb0017421

Analysis and Verification

Granularity of Conflicts and Dependencies in Graph Transformation Systems

Kristopher Born[1], Leen Lambers[2], Daniel Strüber[3], and Gabriele Taentzer[1(✉)]

[1] Philipps-Universität Marburg, Marburg, Germany
{born,taentzer}@informatik.uni-marburg.de
[2] Hasso-Plattner-Institut, Potsdam, Germany
leen.lambers@hpi.de
[3] Universität Koblenz-Landau, Koblenz, Germany
strueber@uni-koblenz.de

Abstract. Conflict and dependency analysis (CDA) is a static analysis for the detection of conflicting and dependent rule applications in a graph transformation system. The state-of-the-art CDA technique, critical pair analysis, provides its users the benefits of completeness, i.e., its output contains a precise representation of each potential conflict and dependency in a minimal context, called *critical pair*. Yet, user feedback has shown that critical pairs can be hard to understand; users are interested in core information about conflicts and dependencies occurring in various combinations. In this paper, we investigate the granularity of conflicts and dependencies in graph transformation systems. We introduce a variety of new concepts on different granularity levels: We start with *conflict atoms*, representing individual graph elements as smallest building bricks that may cause a conflict. We show that each conflict atom can be extended to at least one *conflict reason* and, conversely, each conflict reason is covered by atoms. Moreover, we relate conflict atoms to *minimal conflict reasons*, representing smallest element sets to be overlapped in order to obtain a pair of conflicting transformations. We show how conflict reasons are related to critical pairs. Finally, we introduce dual concepts for dependency analysis. As we discuss in a running example, our concepts pave the way for an improved CDA technique.

1 Introduction

Graph transformation systems (GTSs) are a fundamental modeling concept with applications in a wide range of domains, including software engineering, mechanical engineering, and chemistry. A GTS comprises a set of transformation rules that are applied in coordination to achieve a higher-level goal. The order of rule applications can either be specified explicitly using a control flow mechanism, or it is given implicitly by causal dependencies of rule applications. In the latter case, conflicts and dependencies affect the control flow. For instance, a rule may delete an element whose existence is required by another rule to modify the graph.

© Springer International Publishing AG 2017
J. de Lara and D. Plump (Eds.): ICGT 2017, LNCS 10373, pp. 125–141, 2017.
DOI: 10.1007/978-3-319-61470-0_8

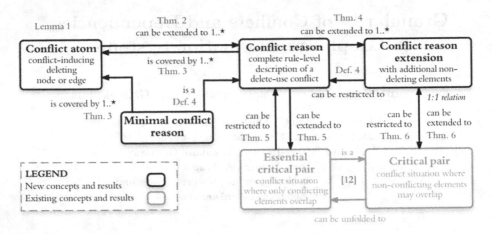

Fig. 1. Inter-relations of new and existing conflict notions

To better understand the implicit control flow of a GTS, one needs to analyze the potential conflicts and dependencies of its rule applications. Conflict and dependency analysis (CDA) is a static analysis for the detection of such conflicts and dependencies. An important CDA technique is *critical pair analysis* [1,2], which has been used in the literature to detect conflicting functional requirements [3], feature interactions [4], conflicting and dependent change operations for process models [5], causal dependencies of aspects in aspect modeling [6], potential conflicts and dependencies between refactorings [7,8], and to validate service-oriented architectures [9].

In these applications, there are generally two possible usage scenarios for CDA: First, the user may start with a list of expected conflicts and dependencies that are supposed to occur. CDA is then used to determine if the expected conflicts and dependencies in fact arise, and/or if there are any unexpected conflicts and dependencies. Violations of expectations signify potential errors in the rule specifications, and can be used for debugging [10]. Second, the user may want to improve their transformation system to reduce conflicts and dependencies, so that rules can be applied independently, e.g., to enable a collaborative modeling process based on edit operation rules [11]. In this case, conflicts and dependencies reported by CDA can be used to identify required modifications. In both cases, users need to inspect conflicts or dependencies to pinpoint their root causes.

To support users during this task, in this work, we lay the basis for a refined CDA technique, distinguishing a variety of new concepts to describe conflicts and dependencies between rules. Our investigation is guided by the notion of *granularity*, and, building on the existing theory for algebraic graph transformation, focuses on *delete-use*-conflicts. We introduce a variety of new concepts and their relations as summarized in Fig. 1. First, we introduce *conflict atoms*, i.e., single graph elements causing conflicts, to represent smallest entities of conflicts.

Each conflict atom can be embedded into the *conflict reason* of a pair of conflicting rules, while each such conflict reason is fully covered by conflict atoms. A conflict reason comprises all elements being deleted by the first and required by the second rule of the considered rule pair. Conflict reasons correspond to essential critical pairs as introduced in previous work [12]. A special type of conflict reasons are *minimal conflict reasons*, representing conflicting graphs and embeddings that are minimal in the sense that they comprise smallest sets of elements required to yield a valid pair of conflicting transformations. Fourth and finally, conflict reasons can be augmented to conflict reason extensions, which have a one-to-one relationship with the notion of critical pairs [1]. Conflict atoms and minimal conflict reasons are more coarse-grained in the sense that they generally represent a larger number of potential conflicts while abstracting away many details of these conflicts, whereas conflict reasons and conflict reason extensions are more fine-grained since they describe conflicts more precisely.

With this contribution, we aim to improve on the state-of-the-art CDA technique, critical pair analysis (CPA) [1,2], by offering improved support for cases where the CPA results did not match the user expectations. In CPA, all potential conflicts and dependencies that can occur when applying two rules are displayed in a minimal context. Confidence in CPA is established by positive fundamental results: via the Completeness Theorem, there exists a critical pair for each conflict, representing this conflict in a minimal context. However, experiences with the CPA indicate two drawbacks: (i) understanding the identified critical pairs can be a challenging task since they display too much information (i.e., they are too fine-grained), (ii) calculating the results can be computationally expensive. Our investigation provides the basis for a solution to compute and report potential conflicts on a level of detail being suitable for the task at hand.

In this paper, we investigate the granularity of conflict and dependencies in GTSs. Specifically, we make three contributions.

- We present a conceptual consideration of conflicts in GTSs, based on the notion of granularity, and focusing on delete-use-conflicts.
- We introduce a variety of formal results for interrelating the new concepts with each other and with the existing concepts. In particular, we relate the new concepts to the well-known conflict concepts of essential and regular critical pairs.
- We discuss how these concepts and results can be transferred to dependencies in a straight-forward manner. In particular, we introduce dependency atoms and reasons, the dual concepts to those introduced for conflict analysis.

The rest of this paper is structured as follows: In Sect. 2, we recall graph transformation concepts and conflict notions from the literature. In Sect. 3, we present the new concepts and formal results. Finally, we compare with related work and conclude in Sect. 4.

2   Preliminaries

As a prerequisite for our new analysis of conflicts and dependencies, we recall the double-pushout approach to graph transformation as presented in [2]. Furthermore, we reconsider two notions of conflicting transformation and their equivalence as shown in [12].

2.1   Graph Transformation: Double-Pushout Approach

Throughout this paper we consider graphs and graph morphisms as presented in [2]; since most of the definitions and results are given in a category-theoretical way, the extension to e.g. typed, attributed graphs [2] is prepared, but up to future work.

Graph transformation is the rule-based modification of graphs. A *rule* mainly consists of two graphs: L is the left-hand side (LHS) of the rule representing a pattern that has to be found to apply the rule. After the rule application, a pattern equal to R, the right-hand side (RHS), has been created. The intersection $K = L \cap R$ is the graph part that is not changed, the graph part that is to be deleted is defined by $L \backslash (L \cap R)$, while $R \backslash (L \cap R)$ defines the graph part to be created. Throughout this paper we consider a graph transformation system just as a set of rules.

A *graph transformation step* $G \overset{m,r}{\Longrightarrow} H$ between two graphs G and H is defined by first finding a graph morphism[1] m of the LHS L of rule r into G such that m is injective, and second by constructing H in two passes: (1) build $D := G \backslash m(L \backslash K)$, i.e., erase all graph elements that are to be deleted; (2) construct $H := D \cup m'(R \backslash K)$ such that a new copy of all graph elements that are to be created is added. It has been shown for graphs and graph transformations that r is applicable at m iff m fulfills the *gluing condition* [2]. In that case, m is called a *match*. For injective morphisms as we use them here, the gluing condition reduces to the *dangling condition*. It is satisfied if all adjacent graph edges of a graph node to be deleted are deleted as well, such that D becomes a graph. Injective matches are usually sufficient in applications and w.r.t. our work here, they allow to explain constructions much easier than for general matches.

Definition 1 (Rule and transformation). *A rule r is defined by $r = (L \hookleftarrow K \hookrightarrow R)$ with L, K, and R being graphs connected by two graph inclusions. A (direct) transformation $G \overset{m,r}{\Longrightarrow} H$ which applies rule r to a graph G consists of two pushouts as depicted below. Morphism $m : L \to G$ is injective and is called match. Rule r is applicable at match m if there exists a graph D such that $(PO1)$ is a pushout.*

---

[1] A morphism between two graphs consists of two mappings between their nodes and edges being both structure-preserving w.r.t. source and target functions. Note that we denote inclusions by $\hookrightarrow$ and all other morphisms by $\to$.

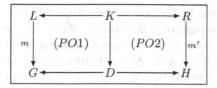

Example 1. Refactoring is a generally acknowledged technique to improve the design of an object-oriented system [13]. To achieve a larger improvement there is typically a sequence of refactorings required. Due to implicit conflicts and dependencies that may occur between refactorings, it is not always easy for developers to determine which refactorings to use and in which order to apply them. To this aim, CDA can support the developer in finding out if there are conflicts or dependencies at all and, if this is the case, in understanding them.

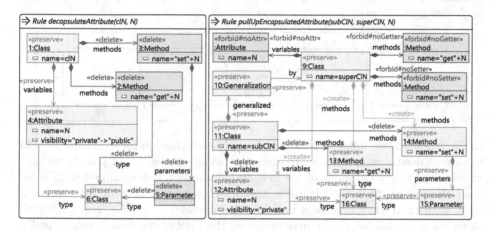

Fig. 2. Refactoring rules *decapsulateAttribute* and *pullUpEncapsulatedAttribute*.

Assuming graphs that model the class design of software systems, we consider Fig. 2 for two class model refactorings being specified as graph-based transformation rules. Rules are depicted in an integrated form where annotations specify which graph elements are deleted, preserved, and created. While the preserved and deleted elements form the LHS of a rule, the preserved and created elements form its RHS. Moreover, negative application conditions specify graph elements that are forbidden when applying a rule. Rule *decapsulateAttribute* removes getter and setter methods for a given attribute, thus inverting the well-known *encapsulate attribute* refactoring. Rule *pullUpEncapsulatedAttribute* takes an attribute with its getter and setter methods and moves them to a superclass if there are not already equally named elements.

2.2   Conflicting Transformations

In this subsection, we recall the essence of conflicting transformations. We concentrate on delete-use conflicts which means that the first rule application deletes graph items that are used by the second rule application. In the literature, there are two different definitions for delete-use conflicts. We recall these definitions and a theorem which shows the equivalence between these two.

The first definition [2] of a delete-use conflict states that the match for the second transformation cannot be found anymore after applying the first transformation. Note that we do not consider delete-use conflicts of the second transformation on the first one explicitly. To get also those ones, we simply consider the inverse pair of transformations.

Definition 2 (Delete-use conflict). *Given a pair of direct transformations* $(t_1, t_2) = (H_1 \overset{m_1, r_1}{\Longleftarrow} G \overset{m_2, r_2}{\Longrightarrow} H_2)$ *applying rules* $r_1 : L_1 \overset{le_1}{\hookleftarrow} K_1 \overset{ri_1}{\hookrightarrow} R_1$ *and* $r_2 : L_2 \overset{le_2}{\hookleftarrow} K_2 \overset{ri_2}{\hookrightarrow} R_2$ *as depicted below. Transformation* t_1 *causes a delete-use conflict on transformation* t_2 *if there does not exist a morphism* $x : L_2 \to D_1$ *such that* $g_1 \circ x = m_2$.

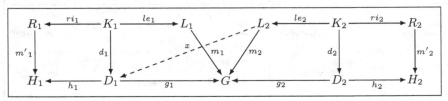

In the following, we consider an alternative characterization for a transformation to cause a delete-use conflict on another one (as introduced in [12]). It states that at least one deleted element of the first transformation is overlapped with some used element of the second transformation. This overlap is formally expressed by a span of graph morphisms between the minimal graph C_1, containing all elements to be deleted by the first rule, and the LHS of the second rule (Fig. 3). In particular, we use an initial pushout construction [2] over the left-hand side morphism of the rule to compute the *boundary graph* B_1 consisting of all nodes needed to make $L_1 \backslash K_1$ a graph and the *context graph* $C_1 := L_1 \backslash (K_1 \backslash B_1)$. We say that the nodes in B_1 are *boundary nodes*. The equivalence of these two conflict notions is recalled in the following theorem.

Theorem 1 (Delete-use conflict characterization). *Given a pair of transformations* $(t_1, t_2) = (H_1 \overset{m_1, r_1}{\Longleftarrow} G \overset{m_2, r_2}{\Longrightarrow} H_2)$ *via rules* $r_1 : L_1 \overset{le_1}{\hookleftarrow} K_1 \overset{ri_1}{\hookrightarrow} R_1$ *and* $r_2 : L_2 \overset{le_2}{\hookleftarrow} K_2 \overset{ri_2}{\hookrightarrow} R_2$, *the initial pushout (1) for* $K_1 \overset{le_1}{\hookrightarrow} L_1$, *and the pullback (2) of* $(m_1 \circ c_1, m_2)$ *in Fig. 2 yielding the span* $s_1 : C_1 \overset{o_1}{\hookleftarrow} S_1 \overset{q_{12}}{\rightrightarrows} L_2$, *then the following equivalence holds:* t_1 *causes a delete-use conflict on* t_2 *according to Definition 2 iff* $s_1 : C_1 \overset{o_1}{\hookleftarrow} S_1 \overset{q_{12}}{\rightrightarrows} L_2$ *satisfies the* conflict condition *i.e. there does not exist any morphism* $x : S_1 \to B_1$ *such that* $b_1 \circ x = o_1$.

In the rest of the paper we merely consider delete-use conflicts such that in the following we abbreviate *delete-use conflict* with *conflict*.

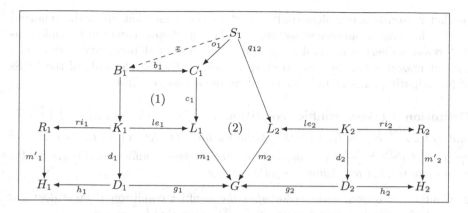

Fig. 3. Delete-use conflict characterization for transformations

3   The Granularity of Conflicts and Dependencies

So far, a conflict between two transformations has always been considered as a whole. In the following, we investigate new notions of conflicting rules presenting them on different levels of granularity. Our intention is the possibility to gradually introduce users to conflicts. Starting with a coarse-grained conflict description in the form of conflict atoms, more information is gradually added until we arrive at the fine-grained representation of conflicts by critical pairs (as e.g. presented in [2]), representing each pair of conflicting transformations in a minimal context. Following this path we introduce several new concepts for conflicting rules and show their interrelations as well as their relations to (essential) critical pairs. Finally, we sketch dual concepts for dependencies.

3.1   Conflicting Rules: Considering Different Granularity Levels

Now, we lift our conflict considerations from transformations to the rule level, i.e., we consider conflicting rules. Two rules are in conflict if there is a pair of conflicting transformations applying these rules. According to Theorem 1 there is a span between these rules specifying the conflict reasons or at least parts of it. In the following, we will concentrate on these spans and distinguish several forms of spans showing conflict reasons in different granularity.

We start focusing on minimal building bricks, called conflict atoms. In particular, we consider a conflict atom to be a minimal sub-graph of C_1 which can be embedded into L_2 but not into B_1 (*conflict and minimality conditions*). Moreover, a pair of direct transformations needs to exist for which the match morphisms overlap on the conflict atom (*transformation condition*). Note that, in general, the matches of this pair of transformations may overlap also in graph elements not contained in the conflict atom. Hence, such a pair of transformations may be chosen flexibly, it need not show a conflict in a minimal context as critical pairs do. While conflict atoms describe the smallest conflict parts, a conflict reason is a complete conflict part in the sense that all in the reported conflict involved atoms are subsumed by it (*completeness condition*). While conflict reasons overlap in conflicting graph elements and boundary nodes only, conflict reason extensions may overlap in non-conflicting elements of the LHSs of participating rules as well (*extended completeness condition*).

Definition 3 (Basic conflict conditions). *Given rules* $r_1 : L_1 \overset{le_1}{\hookleftarrow} K_1 \overset{ri_1}{\hookrightarrow} R_1$ *and* $r_2 : L_2 \overset{le_2}{\hookleftarrow} K_2 \overset{ri_2}{\hookrightarrow} R_2$ *with the initial pushout (1) for* $K_1 \overset{le_1}{\hookrightarrow} L_1$ *as well as a span* $s_1 : C_1 \overset{o_1}{\hookleftarrow} S_1 \overset{q_{12}}{\hookrightarrow} L_2$ *as depicted in Fig. 3, basic conflict conditions for the span* s_1 *of* (r_1, r_2) *are defined as follows:*

1. Conflict condition: *Span* s_1 *satisfies the* conflict condition *if there does not exist any injective morphism* $x : S_1 \to B_1$ *such that* $b_1 \circ x = o_1$.
2. Transformation condition: *Span* s_1 *satisfies the* transformation condition *if there is a pair of transformations* $(t_1, t_2) = (H_1 \overset{m_1, r_1}{\Longleftarrow} G \overset{m_2, r_2}{\Longrightarrow} H_2)$ *via* (r_1, r_2) *with* $m_1(c_1(o_1(S_1))) = m_2(q_{12}(S_1))$ *(i.e. (2) is commuting in Fig. 3).*
3. Completeness condition: *Span* s_1 *satisfies the* completeness condition *if there is a pair of transformations* $(t_1, t_2) = (H_1 \overset{m_1, r_1}{\Longleftarrow} G \overset{m_2, r_2}{\Longrightarrow} H_2)$ *via* (r_1, r_2) *such that (2) is the pullback of* $(m_1 \circ c_1, m_2)$ *in Fig. 3.*
4. Minimality condition: *A span* $s'_1 : C_1 \overset{o'_1}{\hookleftarrow} S'_1 \overset{q'_{12}}{\hookrightarrow} L_2$ *can be embedded into span* s_1 *if there is an injective morphism* $e : S'_1 \to S_1$, *called* embedding morphism, *such that* $o_1 \circ e = o'_1$ *and* $q_{12} \circ e = q'_{12}$. *If* e *is an isomorphism, then we say that the spans* s_1 *and* s'_1 *are* isomorphic. *(See (3) and (4) in Fig. 4.) Span* s_1 *satisfies the* minimality condition *w.r.t. a set* SP *of spans if any* $s'_1 \in SP$ *that can be embedded into* s_1 *is isomorphic to* s_1.

Finally, span $s : L_1 \overset{a_1}{\hookleftarrow} S \overset{b_2}{\hookrightarrow} L_2$ fulfills the *extended completeness condition* if there is a pair of transformations $(t_1, t_2) = (H_1 \overset{m_1, r_1}{\Longleftarrow} G \overset{m_2, r_2}{\Longrightarrow} H_2)$ via (r_1, r_2) such that s arises from the pullback of (m_1, m_2) in the figure on the right.

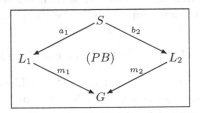

In the following, we define the building bricks of conflicts. The most basic notion to describe a conflict between two rules is that of a *conflict part*. Conflict parts may not describe the whole conflict between two rules. The smallest conflict parts are *conflict atoms*. If a conflict part describes a complete conflict, it is called *conflict reason*.

Definition 4 (Conflict notions for rules). *Let the rules* $r_1 : L_1 \overset{le_1}{\hookleftarrow} K_1 \overset{ri_1}{\hookrightarrow} R_1$ *and* $r_2 : L_2 \overset{le_2}{\hookleftarrow} K_2 \overset{ri_2}{\hookrightarrow} R_2$ *with initial pushout (1) for* $K_1 \overset{le_1}{\hookrightarrow} L_1$ *and a span* $s_1 : C_1 \overset{o_1}{\hookleftarrow} S_1 \overset{q_{12}}{\hookrightarrow} L_2$ *as depicted in Fig. 3, be given.*

1. *Span* s_1 *is called* conflict part candidate *for the pair of rules* (r_1, r_2) *if it satisfies the conflict condition. Graph* S_1 *is called the* conflict graph *of* s_1.

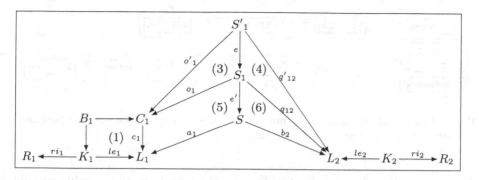

Fig. 4. Illustrating span embeddings

2. *A conflict part candidate s_1 for (r_1, r_2) is a* conflict part *for (r_1, r_2) if s_1 fulfills the transformation condition.*
3. *A conflict part candidate s_1 for (r_1, r_2) is a* conflict atom candidate *for (r_1, r_2) if it fulfills the minimality condition w.r.t. the set of all conflict part candidates for (r_1, r_2).*
4. *A conflict atom candidate s_1 for (r_1, r_2) is a* conflict atom *for (r_1, r_2) if s_1 fulfills the transformation condition.*
5. *A conflict part s_1 for (r_1, r_2) is a* conflict reason *for (r_1, r_2) if s_1 fulfills the completeness condition.*
6. *A conflict reason s_1 for (r_1, r_2) is* minimal *if it fulfills the minimality condition w.r.t. the set of all conflict reasons for (r_1, r_2).*
7. *Span $s : L_1 \xleftarrow{a_1} S \xrightarrow{b_2} L_2$ is a* conflict reason extension *for (r_1, r_2) if it fulfills the extended completeness condition and if there exists a conflict reason s_1 for (r_1, r_2) with $e' : S_1 \to S$ a so-called embedding morphism being injective such that (5) and (6) in Fig. 4 commute. If the latter is the case, we say that s_1 can be embedded via e' into s.*

Note that a conflict part fulfilling the minimality condition is a conflict atom.

Example 2 (Conflict atoms and minimal conflict reasons). Our two example rules in Fig. 2 lead to four pairs of rule combinations to analyze regarding potential conflicts. To discuss the afore introduced building bricks of conflicts we focus on conflicts that may arise by the rule pair (*decapsulateAttribute*, *pullUp-EncapsulatedAttribute*), that means by applying the rule *decapsulateAttribute* and making rule *pullUpEncapsulatedAttribute* inapplicable. Since we do not consider attributes and NACs explicitly in this paper, we neglect them within our conflict analysis. Since these features may restrict rule applications, this decision might lead to an over-approximation of potential conflicts.

The root cause of potential conflicts are the three nodes *2:Method*, *3:Method* and *5:Parameter* to be deleted by rule *decapsulateAttribute*. Nodes of the same type are to be used in rule *pullUpEncapsulatedAttribute*. *Method*-nodes are to be deleted twice by rule *decapsulateAttribute* as well as to be used twice in rule *pullUpEncapsulatedAttribute*. Building all combinations this leads to four

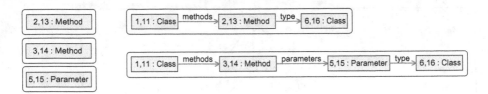

Fig. 5. Conflict atoms (left) and minimal conflict reasons (right) of rule pair (*decapsulateAttribute, pullUpEncapsulatedAttribute*)

different conflict atom candidates. Due to the transformation condition, only two of them are conflict atoms: *2,13:Method* and *3,14:Method*, as depicted in Fig. 5 on the left. A further conflict atom is *5,15:Parameter* which is deleted by *decapsulateAttribute* and used by *pullUpEncapsulatedAttribute*. Note that the span notation is rather compact here: Identifying node numbers of rules are used to indicate the mappings of the atom graph into rule graphs. The three conflict atoms are embedded into two *minimal conflict reasons*. Conflict atom *2,13:Method* and the nodes *1,11:Class* and *6,16:Class* are involved within the first *minimal conflict reason*. The remaining two conflict atoms, *3,14:Method* and *5,15:Parameter* can only be covered by a common *minimal conflict reason* due to the completeness condition. This second *minimal conflict reason* also involves nodes *1,11:Class* and *6,16:Class*. These results provide a concise overview on the root causes of the potential conflicts. The three conflict atoms outline the elements responsible for conflicts and the minimal conflict reasons put them into context to their adjacent nodes.

Remark 1 (Conflict reasons for rules). In [12], a conflict reason is defined for a given pair of direct transformations (t_1, t_2). Here, we lift the notion of conflict reason to a given pair of rules and relate it with the notion of conflict part. In fact, the above definition of conflict reason for rules requires that at least one pair of transformations exists with exactly this conflict part as conflict reason. While a pair of conflicting transformations has a unique conflict reason, two rules may be related by multiple conflict reasons. Note, moreover, that our conflict reason notion for rules is not completely analogous to the notion of conflict reason for transformations in [12]. It would be analogous if we considered conflict reasons where both rules are responsible together for delete-use-conflicts. Since such conflict reasons would be constructed from the other ones, and since we aim for compact representations of conflicts, we opted for not including this case separately.

Table 1 provides a conflict notion overview and basic conditions.

3.2   Relations Between Conflict Notions of Different Granularities

The subsequent results clarify the main interrelations between the new description forms for conflicting rules. All proofs of new results can be found in [14].

Table 1. Overview of conflict concepts

Basic condition/conflict concept	Conflict condition	Transf. condition	Compl. condition	Minimality condition
Conflict part candidate	x			
Conflict part	x	x		
Conflict atom candidate	x			x
Conflict atom	x	x		x
Conflict reason	x	x	x	
Min. conflict reason	x	x	x	x

In the following extension theorem we state that each conflict part can be extended to a conflict reason. As a special case, it follows automatically that each conflict atom (being a special conflict part) can be extended to a conflict reason.

Theorem 2 (Extension of conflict part to reason). *Given a conflict part* $s_1 : C_1 \overset{o_1}{\hookleftarrow} S_1 \overset{q_{12}}{\rightarrowtail} L_2$ *for rule pair* $(r_1 : L_1 \overset{le_1}{\hookleftarrow} K_1 \overset{ri_1}{\hookrightarrow} R_1, r_2 : L_2 \overset{le_2}{\hookleftarrow} K_2 \overset{ri_2}{\hookrightarrow} R_2)$, *there is a conflict reason* $s'_1 : C_1 \overset{o'_1}{\hookleftarrow} S'_1 \overset{q'_{12}}{\rightarrowtail} L_2$ *for* (r_1, r_2) *such that the conflict part* s_1 *can be embedded into it.*

The following lemma gives a more constructive characterization of conflict atom candidates compared to their introduction in Definition 4. This result helps us to characterize conflict atom candidates for a given pair of rules. Candidates are either nodes deleted by rule r_1 and used by rule r_2 or edges deleted by r_1 and used by r_2 if their incident nodes are preserved by r_1. Edges with at least one incident deleted node are not considered as atom candidates since their deletion is caused by node deletions.

Lemma 1 (Conflict atom candidate characterization). *A conflict atom candidate* $s_1 : C_1 \overset{o_1}{\hookleftarrow} S_1 \overset{q_{12}}{\rightarrowtail} L_2$ *for rules* $(r_1 : L_1 \overset{le_1}{\hookleftarrow} K_1 \overset{ri_1}{\hookrightarrow} R_1, r_2 : L_2 \overset{le_2}{\hookleftarrow} K_2 \overset{ri_2}{\hookrightarrow} R_2)$ *has a conflict graph* S_1 *either consisting of a node* v *s.t.* $o_1(v) \in C_1 \backslash B_1$ *or consisting of an edge* e *with its incident nodes* v_1 *and* v_2 *s.t.* $o_1(e) \in C_1 \backslash B_1$ *and* $o_1(v_1), o_1(v_2) \in B_1$.

Note that, for attributed graphs, the edge in a conflict atom may also be an attribute edge. In this case, the conflict atom would describe an attribute change which is in conflict with an attribute use.

The following theorem states that each conflict reason is covered by a unique set of atoms, i.e. all atoms that can be embedded into that conflict reason. With atoms we mean conflict atoms as well as boundary atoms, where a latter one consists merely of a single boundary node. This means that by investigating the set of conflict atoms one gets a complete overview of graph elements that can cause conflicts in a given conflict reason. Moreover, the set of boundary atoms indicates how this conflict reason might be still enlarged with other conflict-inducing edges. Of course, this result also holds for the special case that the conflict reason is minimal.

Definition 5 (Boundary atom). *A span $s_1^b : C_1 \overset{o_1^b}{\hookleftarrow} S_1^b \overset{q_{12}^b}{\rightarrow} L_2$ is a boundary part for rules (r_1, r_2) with initial pushout (1) as in Fig. 3 if there is a morphism $s_B : S_1^b \rightarrow B_1$ such that $b_1 \circ s_B = o_1^b$ and s_1^b fulfills the transformation condition. A non-empty boundary part s_1^b is a boundary atom if it fulfills the minimality condition w.r.t. the set of boundary parts for (r_1, r_2).*

It is straightforward to show that graph S_1^b of a boundary atom consists of exactly one boundary node being the source or target node of an edge that is potentially conflict-inducing.

Theorem 3 (Covering of conflict reasons by atoms). *Given a conflict reason $s_1 : C_1 \overset{o_1}{\hookleftarrow} S_1 \overset{q_{12}}{\rightarrow} L_2$ for rules (r_1, r_2), then the set A of all conflict atoms together with the set A^B of all boundary atoms that can be embedded into s_1 covers s_1, i.e. for each conflict reason $s'_1 : C_1 \overset{o'_1}{\hookleftarrow} S'_1 \overset{q'_{12}}{\rightarrow} L_2$ for (r_1, r_2) that can be embedded into s_1 it holds that, if each atom in $A \cup A^B$ can be embedded into s'_1, then s'_1 is isomorphic to s_1.*

Conflict reason extensions contain all graph elements that overlap in a pair of conflicting transformations, even elements that are not deleted and at the same time used by any of the two participating rules. Hence, a conflict reason extension might show too much information. By definition, for each conflict reason extension, there is a conflict reason which can be embedded into this extension. Hence, an extension can always be restricted to a conflict reason. Vice versa, the following theorem shows that each conflict reason (being defined over C_1 and L_2) can be extended to at least one conflict reason extension (being defined over L_1 and L_2).

Theorem 4 (Extension of conflict reason to conflict reason extension). *Given a conflict reason $s_1 : C_1 \overset{o_1}{\hookleftarrow} S'_1 \overset{q_{12}}{\rightarrow} L_2$ for rules (r_1, r_2), there exists at least one conflict reason extension $s : L_1 \overset{a_1}{\hookleftarrow} S \overset{b_2}{\rightarrow} L_2$ for rules (r_1, r_2) such that s_1 can be embedded into s.*

3.3   Relations of Conflicting Rule Concepts to Critical Pairs

As illustrated in Fig. 1, for each *critical pair*, there exists an essential critical pair that can be embedded into it (see Completeness Theorem 4.1 in [12]). Match pairs of each (essential) critical pair are jointly surjective (according to the minimal context idea). Thus a critical pair might overlap in elements that are just read by both rules and are not boundary nodes, and exactly these overlaps are *unfolded* again in the essential critical pair. This is because the latter overlaps do not contribute to a new kind of conflict. The set of essential critical pairs is thus smaller than the set of critical pairs and, in particular, each essential critical pair is a critical pair (see Fact 3.2 in [12]).

The following two theorems formalize, on the one hand, the relations between conflict reasons for rule pairs as introduced in this paper and essential critical pairs, and on the other hand, the relations between conflict reason extensions and critical pairs. Note that, as explained in Remark 1, there is no 1–1 correspondence of conflict reasons for rules and essential critical pairs, since we abstract from building symmetrical conflict reasons on the rule level for compactness reasons.

Theorem 5 (Essential critical pair and conflict reason). Restriction. *Given an essential critical pair $(t_1, t_2) = (P_1 \overset{m_1,r_1}{\Longleftarrow} K \overset{m_2,r_2}{\Longrightarrow} P_2)$ such that t_1 causes a delete-use conflict on t_2 then the span $s_1 : C_1 \overset{o_1}{\hookleftarrow} S_1 \overset{q_{12}}{\hookrightarrow} L_2$ arising from taking the pullback of $(m_1 \circ c_1, m_2)$ is a conflict reason for (r_1, r_2).*

Extension. *Given a conflict reason $s_1 : C_1 \overset{o_1}{\hookleftarrow} S_1 \overset{q_{12}}{\hookrightarrow} L_2$ for rule pair (r_1, r_2) then there exists an essential critical pair $(t_1, t_2) = (P_1 \overset{m_1,r_1}{\Longleftarrow} K \overset{m_2,r_2}{\Longrightarrow} P_2)$ such that t_1 causes a delete-use conflict on t_2 with the pullback of $(m_1 \circ c_1, m_2)$ being isomorphic to s_1.*

Theorem 6 (Critical pair and conflict reason extension). Restriction. *Given a critical pair $(t_1, t_2) = (P_1 \overset{m_1,r_1}{\Longleftarrow} K \overset{m_2,r_2}{\Longrightarrow} P_2)$ such that t_1 causes a delete-use conflict on t_2 then the span arising from taking the pullback of (m_1, m_2) is a conflict reason extension for (r_1, r_2).*

Extension. *Given a conflict reason extension $s : L_1 \overset{a_1}{\hookleftarrow} S \overset{b_2}{\hookrightarrow} L_2$ for (r_1, r_2) then the cospan arising from building the pushout of (a_1, b_2) defines the matches (m_1, m_2) of a critical pair $(t_1, t_2) = (P_1 \overset{m_1,r_1}{\Longleftarrow} K \overset{m_2,r_2}{\Longrightarrow} P_2)$ such that t_1 causes a delete-use conflict on t_2.*

Bijective correspondence. *The restriction and extension constructions are inverse to each other up to isomorphism.*

Example 3 (Conflict reason extension). Figure 5 focuses on the conflict atoms and minimal conflict reasons of the rule pair (*decapsulateAttribute, pullUpEncapsulatedAttribute*). Figure 6 relates these new conflict notions with the six critical pairs of the considered rule pair. The two *minimal conflict reasons* sufficiently characterize the overlap in the results 3 and 5. Result 1 presents the combination of both *minimal conflict reasons*. Since these results make no use of further overlapping of non-deleting elements they are also conflict reason extensions. Moreover, they correspond to the results of the essential critical pair analysis. *1,11:Class* and *6,16:Class* are two *boundary atoms*. Additional overlapping of the *Attribute*-nodes of both rules in *4,12:Attribute* leads to larger *conflict reason extensions* and to the remaining three results 2, 4, and 6. Adding the remaining elements of the LHS of both rules, we obtain a compact representation of all six critical pairs.

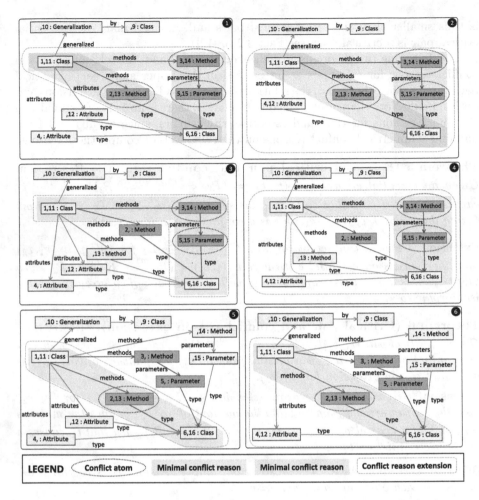

Fig. 6. Representation of six critical pairs arising from the application of rule *decapsulateAttribute* so that rule *pullUpEncapsulatedAttribute* becomes inapplicable; examples of newly introduced conflict notions are indicated.

3.4   Dual Notions for Dependencies

To reason about dependencies of rules and transformations, we consider the dual concepts and results that we get when inverting the left transformation of a conflicting pair. This means that we check if $H_1 \overset{p_1^{-1}, m'_1}{\Longleftarrow} G \overset{p_2, m_2}{\Longrightarrow} H_2$ is parallel dependent, which is equivalent to the sequence $G \overset{p_1, m_1}{\Longrightarrow} H_1 \overset{p_2, m_2}{\Longrightarrow} H_2$ being sequentially dependent. This is possible since a transformation is symmetrically defined by two pushouts. They ensure in particular that morphisms $m : L \to G$ as well as $m' : R \to H$ fulfill the gluing condition.

Dependency parts, atoms, reasons, and reason extensions can be defined analogously to Definition 4. They characterize graph elements being produced by the first rule application and used by the second one. Results presented for conflicts above can be formulated and proven for dependencies in an analogous way.

4 Related Work and Conclusion

The critical pair analysis (CPA) has developed into the standard technique for detecting conflicts and dependencies in graph transformation systems [1] at design time. Originally being developed for term and term graph rewriting [15], it extends the theory of graph transformation and, more generally, of $\mathcal{M}$-adhesive transformation systems [2,16]. The CPA is not only available for plain rules but also for rules with application conditions [17].

In this paper, we lay the basis for a refined analysis of conflicts and dependencies by presenting conflict and dependency notions of different granularity. Furthermore, we investigate their interrelations. The formal consideration shall be used in a new CDA technique where conflict and dependency analysis can go from coarse-grained information about the potential existence of conflicts or dependencies and their main reasons, to fine-grained considerations of conflict and dependency reasons in different settings.

The CPA is offered by the graph transformation tools AGG [18] and Verigraph [19] and the graph-based model transformation tool Henshin [20]. All of them provide the user with a set of (essential) critical pairs for each pair of rules as analysis result. The computation of conflicts and dependencies using the concepted introduced in the present work has been prototypically implemented in Henshin. First tests indicate that our analysis is very fast and yields concise results that are promising to facilitate understandability. However, it is up to future work to further investigate this aspect in a user study.

Currently, we restrict our formal considerations to graphs and graph transformations. Since all main concepts are based on concepts from category theory, our work is prepared to adapt to more sophisticated forms of graphs or graph transformation. Furthermore, it is interesting to adapt the new notions to transformation rules with negative [21] or more complex nested application conditions [17]. Analogously, to handle attributes within conflicts appropriately it is promising to adapt our approach to lazy graph transformations [22] and to come up with a light-weight conflict analysis complementing the work of Deckwerth et al. [23] on conflict detection of edit operations on feature models. They combine CPA with an SMT solver for an improved handling of conflicts based on attribute changes. Performance is still a limiting factor for applying the CPA to large rule sets. A *family-based analysis* based on the unification of multiple similar rules [24] is a promising idea to save redundant computation effort.

Acknowledgements. We wish to thank Jens Kosiol and the anonymous reviewers for their constructive comments. This work was partially funded by the German Research Foundation, Priority Program SPP 1593 "Design for Future – Managed Software Evolution". This research was partially supported by the research project Visual Privacy Management in User Centric Open Environments (supported by the EU's Horizon 2020 programme, Proposal number: 653642).

References

1. Heckel, R., Küster, J.M., Taentzer, G.: Confluence of typed attributed graph transformation systems. In: Corradini, A., Ehrig, H., Kreowski, H.-J., Rozenberg, G. (eds.) ICGT 2002. LNCS, vol. 2505, pp. 161–176. Springer, Heidelberg (2002). doi:10.1007/3-540-45832-8_14
2. Ehrig, H., Ehrig, K., Prange, U., Taentzer, G.: Fundamentals of Algebraic Graph Transformation. Monographs in Theoretical Computer Science. Springer, Heidelberg (2006)
3. Hausmann, J.H., Heckel, R., Taentzer, G.: Detection of conflicting functional requirements in a use case-driven approach: a static analysis technique based on graph transformation. In: Proceedings of the 22nd International Conference on Software Engineering (ICSE), pp. 105–115. ACM (2002)
4. Jayaraman, P., Whittle, J., Elkhodary, A.M., Gomaa, H.: Model composition in product lines and feature interaction detection using critical pair analysis. In: Engels, G., Opdyke, B., Schmidt, D.C., Weil, F. (eds.) MODELS 2007. LNCS, vol. 4735, pp. 151–165. Springer, Heidelberg (2007). doi:10.1007/978-3-540-75209-7_11
5. Küster, J.M., Gerth, C., Engels, G.: Dependent and conflicting change operations of process models. In: Paige, R.F., Hartman, A., Rensink, A. (eds.) ECMDA-FA 2009. LNCS, vol. 5562, pp. 158–173. Springer, Heidelberg (2009). doi:10.1007/978-3-642-02674-4_12
6. Mehner-Heindl, K., Monga, M., Taentzer, G.: Analysis of aspect-oriented models using graph transformation systems. In: Moreira, A., Chitchyan, R., Araújo, J., Rashid, A. (eds.) Aspect-Oriented Requirements Engineering, pp. 243–270. Springer, Heidelberg (2013)
7. Mens, T., Straeten, R., D'Hondt, M.: Detecting and resolving model inconsistencies using transformation dependency analysis. In: Nierstrasz, O., Whittle, J., Harel, D., Reggio, G. (eds.) MODELS 2006. LNCS, vol. 4199, pp. 200–214. Springer, Heidelberg (2006). doi:10.1007/11880240_15
8. Mens, T., Taentzer, G., Runge, O.: Analysing refactoring dependencies using graph transformation. Softw. Syst. Model. 6(3), 269–285 (2007)
9. Baresi, L., Heckel, R., Thöne, S., Varró, D.: Modeling and validation of service-oriented architectures: application vs. style. In: ACM SIGSOFT Symposium on Foundations of Software Engineering Held Jointly with 9th European Software Engineering Conference, pp. 68–77. ACM (2003)
10. Ermel, C., Gall, J., Lambers, L., Taentzer, G.: Modeling with plausibility checking: inspecting favorable and critical signs for consistency between control flow and functional behavior. In: Giannakopoulou, D., Orejas, F. (eds.) FASE 2011. LNCS, vol. 6603, pp. 156–170. Springer, Heidelberg (2011). doi:10.1007/978-3-642-19811-3_12
11. Strüber, D., Taentzer, G., Jurack, S., Schäfer, T.: Towards a distributed modeling process based on composite models. In: Cortellessa, V., Varró, D. (eds.) FASE 2013. LNCS, vol. 7793, pp. 6–20. Springer, Heidelberg (2013). doi:10.1007/978-3-642-37057-1_2
12. Lambers, L., Ehrig, H., Orejas, F.: Efficient conflict detection in graph transformation systems by essential critical pairs. Electr. Notes Theor. Comput. Sci. 211, 17–26 (2008)
13. Fowler, M.: Refactoring: Improving the Design of Existing Code. Addison-Wesley, Boston (1999)

14. Born, K., Lambers, L., Strüber, D., Taentzer, G.: Granularity of conflicts
    and dependencies in graph transformation systems: extended version, Philipps-
    Universität Marburg, Technical Report (2017). www.uni-marburg.de/fb12/swt/
    research/publications
15. Plump, D.: Critical pairs in term graph rewriting. In: Prívara, I., Rovan, B.,
    Ruzička, P. (eds.) MFCS 1994. LNCS, vol. 841, pp. 556–566. Springer, Heidelberg
    (1994). doi:10.1007/3-540-58338-6_102
16. Ehrig, H., Padberg, J., Prange, U., Habel, A.: Adhesive high-level replacement
    systems: a new categorical framework for graph transformation. Fundam. Inform.
    74(1), 1–29 (2006)
17. Ehrig, H., Golas, U., Habel, A., Lambers, L., Orejas, F.: $\mathcal{M}$-adhesive trans-
    formation systems with nested application conditions. part 2: embedding, crit-
    ical pairs and local confluence. Fundam. Inform. **118**(1–2), 35–63 (2012).
    http://dx.doi.org/10.3233/FI-2012-705
18. Taentzer, G.: AGG: a graph transformation environment for modeling and val-
    idation of software. In: Pfaltz, J.L., Nagl, M., Böhlen, B. (eds.) AGTIVE
    2003. LNCS, vol. 3062, pp. 446–453. Springer, Heidelberg (2004). doi:10.1007/
    978-3-540-25959-6_35
19. Verigraph. https://github.com/Verites/verigraph
20. Arendt, T., Biermann, E., Jurack, S., Krause, C., Taentzer, G.: Henshin: advanced
    concepts and tools for in-place EMF model transformations. In: Petriu, D.C.,
    Rouquette, N., Haugen, Ø. (eds.) MODELS 2010. LNCS, vol. 6394, pp. 121–135.
    Springer, Heidelberg (2010). doi:10.1007/978-3-642-16145-2_9
21. Lambers, L.: Certifying rule-based models using graph transformation, Ph.D. dis-
    sertation, Berlin Institute of Technology (2010)
22. Orejas, F., Lambers, L.: Lazy graph transformation. Fundam. Inform. **118**(1–2),
    65–96 (2012)
23. Deckwerth, F., Kulcsár, G., Lochau, M., Varró, G., Schürr, A.: Conflict detection
    for edits on extended feature models using symbolic graph transformation. In:
    International Workshop on Formal Methods and Analysis in Software Product
    Line Engineering. EPTCS, vol. 206, pp. 17–31 (2016)
24. Strüber, D., Rubin, J., Arendt, T., Chechik, M., Taentzer, G., Plöger, J.: *Rule-*
    Merger: automatic construction of variability-based model transformation rules.
    In: Stevens, P., Wąsowski, A. (eds.) FASE 2016. LNCS, vol. 9633, pp. 122–140.
    Springer, Heidelberg (2016). doi:10.1007/978-3-662-49665-7_8

k-Inductive Invariant Checking
for Graph Transformation Systems

Johannes Dyck$^{(\boxtimes)}$ and Holger Giese

Hasso Plattner Institute, University of Potsdam, Potsdam, Germany
{Johannes.Dyck,Holger.Giese}@hpi.de

Abstract. While offering significant expressive power, graph transformation systems often come with rather limited capabilities for automated analysis, particularly if systems with many possible initial graphs and large or infinite state spaces are concerned. One approach that tries to overcome these limitations is inductive invariant checking. However, the verification of inductive invariants often requires extensive knowledge about the system in question and faces the approach-inherent challenges of locality and lack of context.

To address that, this paper discusses k-inductive invariant checking for graph transformation systems as a generalization of inductive invariants. The additional context acquired by taking multiple (k) steps into account is the key difference to inductive invariant checking and is often enough to establish the desired invariants without requiring the iterative development of additional properties.

To analyze possibly infinite systems in a finite fashion, we introduce a symbolic encoding for transformation traces using a restricted form of nested application conditions. As its central contribution, this paper then presents a formal approach and algorithm to verify graph constraints as k-inductive invariants. We prove the approach's correctness and demonstrate its applicability by means of several examples evaluated with a prototypical implementation of our algorithm.

1 Introduction

The expressive power of graph transformation systems often leads to rather limited capabilities for automated analysis, particularly if systems with many initial graphs and large or infinite state spaces are concerned. Model checkers can typically only be employed for the analysis of graph transformation systems with a finite state space of moderate size (e.g., [9,14]). Other fully automatic approaches that can handle infinite state spaces by abstraction [2,3,11,12,16] are limited in their expressiveness, supporting only limited forms of negative application conditions at most. In some cases, additional limitations concerning the graphs of the state space apply (cf. [2]). In contrast to that, the SeekSat/ProCon tool [10,13] is able to prove correctness of graph programs with respect to pre- and

This work was partially developed in the course of the project Correct Model Transformations II (GI 765/1-2), which is funded by the Deutsche Forschungsgemeinschaft.

© Springer International Publishing AG 2017
J. de Lara and D. Plump (Eds.): ICGT 2017, LNCS 10373, pp. 142–158, 2017.
DOI: 10.1007/978-3-319-61470-0_9

postconditions specified as nested graph constraints; however, it may require too expensive (cf. [5]) or infeasible computations.

One direction that tries to overcome these limitations is the automated verification of *inductive invariants* (cf. our own work [1,5]), where we analyze the capability of system behavior (captured by a number of graph rules) to preserve or violate desired properties (captured by graph constraints) as inductive invariants. However, the technique faces the approach-inherent challenges of locality and lack of context information. The analysis of single transformation steps does not take the broader context, prior rule applications, or the state space into account, which is both the primary objective and a main challenge of the approach. Hence, in order to develop successfully verifiable inductive invariants (if the system is, indeed, safe) or establish meaningful counterexamples (if the system is not), additional knowledge encoded by additional properties may be required and often has to be accumulated by an iterative and manual procedure.

Therefore, this paper applies the notion of k-*induction* [15] to graph transformation systems by extending our previous work in [1,5]. In particular, k-inductive invariants are a generalization of inductive invariants; conversely, the latter are a special case of the former for $k = 1$. Our approach takes paths of length k into account [15]: a k-inductive invariant is a property whose validity in a path of length $k - 1$ implies its validity in the subsequent step. By analyzing system behavior over multiple transformation steps, more context information is available and the resulting analysis will be more precise. While the idea of k-induction has been successfully employed in the field of software verification [4], to the best of our knowledge, no approach to automatically check k-inductive invariants for graph transformation systems has been developed so far.

In order to analyze possibly infinite systems in a finite fashion, we first introduce a symbolic encoding for transformation traces. Our main contribution is a formal approach and algorithm to verify a restricted form of graph constraints as k-inductive invariants. We prove the approach's correctness and demonstrate its applicability by means of several examples evaluated with a prototypical implementation of our algorithm. While our technique takes care of the inductive step (verifying the k-inductive invariant), the base of induction for traces of length $k - 1$ from an initial graph is established with the model checker GROOVE [9].

This paper is organized as follows: In Sect. 2, we reintroduce the necessary foundations and our formal model. Section 3 defines our notion of k-inductive invariants and the symbolic encoding. We present our formal approach to k-inductive invariant checking in Sect. 4. In Sect. 5, we evaluate our algorithm and approach, before summarizing our results in Sect. 6. Omitted constructions, proofs, and more details to our examples can be found in the respective sources and in [6], which is an extended version of this paper.

2   Prerequisites

This section cites formal foundations [7,8,10], introduces our running example, and reintroduces the restricted formal model employed in our approach and tool.

2.1   Foundations

The formalism we use (see [7] and our previous paper [5]) considers a *graph* to consist of node and edge sets V and E and source and target functions $s, t : E \to V$ assigning source and target nodes to edges. A *graph morphism* $f : G_1 \to G_2$ for graphs $G_i = (V_i, E_i, s_i, t_i)$ with $i = 1, 2$ consists of two functions mapping nodes and edges $f = (f_V, f_E)$ with $f_V : V_1 \to V_2$ and $f_E : E_1 \to E_2$ that preserve source and target functions. Injective graph morphisms (or *monomorphisms*) are graph morphisms with injective mapping functions and are denoted as $f : G_1 \hookrightarrow G_2$. A *typed graph* G is typed over a special type graph TG by a *typing morphism* $type : G \to TG$; *typed graph morphisms* must preserve the typing morphism.

Additionally, we require (nested) application conditions [8] and (nested) graph constraints [10] to describe more complex conditions over morphisms and graphs, respectively. Here, an application condition (graph constraint) can also be interpreted as describing the set of morphisms (graphs) that satisfy it.

Application conditions (or *nested application conditions*) are inductively defined as in [8]: (1) for every graph P, true is an application condition over P; (2) for every morphism $a : P \hookrightarrow C$ and every application condition ac over C, $\exists(a, ac)$ is an application condition over P. Application conditions can also be extended over boolean combinations: (3) for application conditions ac, ac_i over P (for all index sets I), $\neg ac$ and $\bigwedge_{i \in I} ac_i$ are application conditions over P.

Satisfiability of application conditions is inductively defined as in [8]: (1) every morphism satisfies true; (2) a morphism $g : P \to G$ satisfies $\exists(a, ac)$ over P with $a : P \to C$ if there exists an injective $q : C \hookrightarrow G$ such that $q \circ a = g$ and q satisfies ac. Finally, (3) a morphism $g : P \to G$ satisfies $\neg ac$ over P if g does not satisfy ac and g satisfies $\bigwedge_{i \in I} ac_i$ over P if g satisfies each ac_i $(i \in I)$.

We write $g \vDash ac$ to denote that the morphism g satisfies ac. Two conditions ac and ac' are equivalent ($ac \equiv ac'$), if for all morphisms $g : P \to G$, $g \vDash ac$ if and only if $g \vDash ac'$. Also, $\exists p$ and $\forall(p, ac)$ abbreviate $\exists(p, \text{true})$ and $\neg \exists(p, \neg ac)$.

A *graph constraint* [10] is a condition over the empty graph $\varnothing$. A graph G then satisfies such a condition if the initial morphism $i_G : \varnothing \hookrightarrow G$ satisfies it.

We use graph rules to describe how graphs can be transformed by rule applications. As defined in [8], a *plain rule* $p = (L \hookleftarrow K \hookrightarrow R)$ consists of two injective morphisms $K \hookrightarrow L$ and $K \hookrightarrow R$. L and R are called left- and right-hand side of p, respectively. A *rule* $b = \langle p, ac_L, ac_R \rangle$ consists of a plain rule p and a *left* (*right*) *application condition* ac_L (ac_R) over L (R).

A *transformation* (also [8]) consists of two pushouts (1) and (2) such that $m \vDash ac_L$ and $m' \vDash ac_R$. We write $G \Rightarrow_{b,m,m'} H$ and say that $m : L \to G$ ($m' : R \to H$) is the *match* (*comatch*) of b in G (in H). We write $G \Rightarrow_b H$ to express that there exist m and m' such that $G \Rightarrow_{b,m,m'} H$. For a set of rules $\mathcal{R}$, $G \Rightarrow_{\mathcal{R}} H$ expresses that there exist $b \in \mathcal{R}$ and m, m' such that $G \Rightarrow_{b,m,m'} H$. Also, given a rule $b = \langle (L \hookleftarrow K \hookrightarrow R), ac_L, ac_R \rangle$, its *inverse rule* is denoted as b^{-1} and defined as $b^{-1} = \langle (R \hookleftarrow K \hookrightarrow L), ac_R, ac_L \rangle$.

$$
\begin{array}{ccccc}
ac_L \triangleright L & \xleftarrow{\; l \;} & K & \xrightarrow{\; r \;} & R \triangleleft ac_R \\
\downarrow m & (1) & \downarrow & (2) & \downarrow m' \\
G & \xleftarrow{\; l' \;} & D & \xrightarrow{\; r' \;} & H
\end{array}
$$

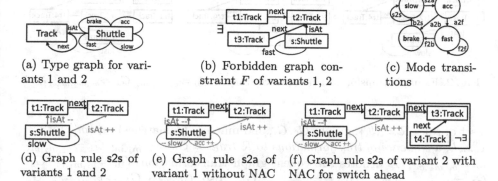

(a) Type graph for variants 1 and 2

(b) Forbidden graph constraint *F* of variants 1, 2

(c) Mode transitions

(d) Graph rule s2s of variants 1 and 2

(e) Graph rule s2a of variant 1 without NAC

(f) Graph rule s2a of variant 2 with NAC for switch ahead

Fig. 1. Example type graph, graph constraint, mode transitions, and rules (Color figure online)

Finally, a *typed graph transformation system GTS* = ($\mathcal{R}$, *TG*) consists of a set of graph rules $\mathcal{R}$ and a type graph *TG* [7].

Example 1 (variant 1; see [6] *for all details).* Our running example is a system where a single shuttle moves on a topology of connected tracks in different speed modes (slow, acc(elerate), fast, and brake), which follow a certain protocol (Fig. 1(c)). The system also has a forbidden property, which describes a shuttle driving on a switch in mode fast. Fig. 1 shows our system modeled as a typed graph transformation system: a type graph (Fig. 1(a)), the forbidden property as a graph constraint (*F*, Fig. 1(b)), and seven rules modeling shuttle movement and driving mode transitions. Two of those rules are depicted: s2s (slow to slow) in Fig. 1(d) and s2a (slow to acc(elerate)) in Fig. 1(e). All other rules (a2b, a2f, f2b, b2s, f2f) function analogously and follow the scheme of s2a or s2s, respectively. Graph rules are pictured in a compact notation: deleted (created) elements are drawn in red (green) and annotated $--$ ($++$); unchanged elements are in black.

This example variant exhibits unsafe behavior in the sense of possible violations of our forbidden property, because rule s2a (Fig. 1(e)) does not prevent the shuttle from accelerating from slow to acc when there is a switch two tracks ahead. Subsequent application of a2f could then lead to the violation. In our second variant, this error has been fixed:

Example 2 (variant 2; see [6] *for all details).* While the type graph, forbidden property, and most rules remain the same, variant 2 modifies variant 1 by extending a number of rules with a left (negative) application condition: s2a as in Fig. 1(f), a2f and f2f in a similar fashion. The application condition is designed to prevent the shuttle from acc(elerating) (and, for a2f and f2f, from driving fast) if a switch is two tracks ahead.

Since *k*-inductive invariant checking considers paths of transformations instead of single steps, we require the notion of transformation sequences. Given

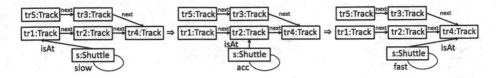

Fig. 2. Example transformation sequence $trans = G_0 \Rightarrow_{b_1,m_1,m_1'} G_1 \Rightarrow_{b_2,m_2,m_2'} G_2$

a set of rules $\mathcal{R} = \{b_i\}$, graphs G_0, G_i, and matches (comatches) m_i (m_i') for $i = 1, ..., k$, a *sequence of transformations to* $\mathcal{R}$ $trans = G_0 \Rightarrow_{b_1,m_1,m_1'} G_1 \Rightarrow_{b_2,m_2,m_2'} ... \Rightarrow_{b_k,m_k,m_k'} G_k$ denotes subsequent graph transformations $G_0 \Rightarrow_{b_1,m_1,m_1'} G_1$, $G_1 \Rightarrow_{b_2,m_2,m_2'} G_2, ..., G_{k-1} \Rightarrow_{b_k,m_k,m_k'} G_k$. Also, $G_0 \Rightarrow_{\mathcal{R}} G_1 \Rightarrow_{\mathcal{R}} ... \Rightarrow_{\mathcal{R}} G_k$ denotes $G_0 \Rightarrow_{\mathcal{R}} G_1, G_1 \Rightarrow_{\mathcal{R}} G_2, ..., G_{k-1} \Rightarrow_{\mathcal{R}} G_k$ and is abbreviated as $G_0 \Rightarrow_{\mathcal{R}}^k G_k$.

We say that a sequence of transformations $trans = G_0 \Rightarrow_{b_1} ... \Rightarrow_{b_k} G_k$ *leads* to a graph constraint F, if $G_k \vDash F$.

Example 3 (transformation sequence). Figure 2 shows a transformation sequence $trans = G_0 \Rightarrow_{b_1,m_1,m_1'} G_1 \Rightarrow_{b_2,m_2,m_2'} G_2$, where b_1 and b_2 are graph rules s2a and a2f from Example 1. Matches and comatches are not depicted. Here, G_2 contains a shuttle on a switch driving in mode fast, which matches our forbidden property F (Fig. 1(b)). Thus, we have $G_2 \vDash F$ and *trans* leads to F. Note that *trans* would not be a valid transformation sequence in variant 2 due to the additional negative application condition preventing the application of s2a (b_1).

An important part of our algorithm is the Shift-construction [8]. Given an application condition ac over a graph P and a morphism $b : P \to P'$, Shift(b, ac) transforms ac via b into an application condition over P' such that, for each morphism $n : P' \hookrightarrow H$, it holds that $n \circ b \vDash ac \Leftrightarrow n \vDash$ Shift(b, ac). Shift is constructed as follows [8]: (1) Shift$(b, \text{true}) = \text{true}$, (2) Shift$(b, \exists(a, ac)) = \bigvee_{(a',b') \in \mathcal{F}} \exists(a', \text{Shift}(b', ac))$ for a non-empty set $\mathcal{F}$ of jointly surjective and injective morphism pairs (a', b') such that $b' \circ a = a' \circ b$, and false, if $\mathcal{F} = \varnothing$, (3) Shift$(b, \neg ac) = \neg$ Shift(b, ac), and (4) Shift$(b, \bigwedge_{i \in I} ac_i) = \bigwedge_{i \in I}$ Shift(b, ac_i).

Furthermore, we use the L-construction [8,10]: Given $b = (L \leftarrow K \to R)$ and a condition ac over R, L(b, ac) transforms ac via b into a condition over L such that, for each transformation $G \Rightarrow_{b,m,m'} H$, we have $m \vDash$ L$(b, ac) \Leftrightarrow m' \vDash ac$. L is inductively defined [8,10]: (1) L$(b, \text{true}) = \text{true}$, (2) L$(b, \exists(a, ac)) = \exists(a', \text{L}(b', ac))$ if $b' = \langle L' \leftarrow K' \to R' \rangle$ constructed via the pushouts (1) and (2) exists and false, otherwise, (3) L$(b, \neg ac) = \neg$ L(b, ac), and (4) L$(b, \bigwedge_{i \in I} ac_i) = \bigwedge_{i \in I}$ L(b, ac_i).

Both Shift and L produce finite results by construction.

2.2   Formal Model

As described in [5], our verification approach and tool impose certain restrictions on rules and properties in order to strike a balance between expressiveness and computational complexity while ensuring termination. In the following, we discuss those restrictions, starting with composed negative application conditions as a restricted form of nested application conditions.

Definition 4 (composed negative application condition [5]**).** *A composed negative application condition is an application condition of the form* $ac = \text{true}$ *or* $ac = \bigwedge_{i \in I} \neg \exists a_i$ *for monomorphisms* a_i. *An individual condition* $\neg \exists a_i$ *is called* negative application condition.

Our properties to be verified as *k*-inductive invariants are described by so-called forbidden patterns, which follow a restricted form of graph constraints.

Definition 5 (pattern [5]**).** *A pattern is a graph constraint of the form* $F = \exists (i_P : \varnothing \hookrightarrow P, ac_P)$, *with* P *being a graph and* ac_P *a composed negative application condition over* P. *A composed forbidden pattern is a graph constraint of the form* $\mathcal{F} = \bigwedge_{i \in I} \neg F_i$ *for some index set* I *and patterns* F_i. *Patterns* F_i *occurring in a composed forbidden pattern are also called* forbidden patterns.

Besides forbidden patterns, we allow our systems to be equipped with *(composed) assumed forbidden patterns*, which are similar in form to (composed) forbidden patterns and which will be explained below (see Example 8).

In order to compare patterns, we reintroduce the notion of pattern implication. Our technique to actually check for pattern implication is described in [5,6]. More general approaches discussing implication of (unrestricted) graph constraints can be found in [10,13].

Definition 6 (implication of patterns [5]**).** *Let* C *and* C' *be two patterns.* C' *implies* C, *denoted* $C' \vDash C$, *if, for all graphs* G, $G \vDash C'$ *implies* $G \vDash C$.

In summary, our formal model is subject to the restrictions listed below [5]. However, only the requirements concerned with left application conditions and graph constraints actually result in a limitation of expressive power [7,8,10].

Morphisms in application conditions (see Sect. 2.1) must be injective.
Left application conditions (see Sect. 2.1) in rules are required to be composed negative application conditions.
Right application conditions (Sect. 2.1) in rules are required to be true.
Rule applicability (see Sect. 2.1) requires injective matches and comatches.
Graph constraints must be patterns (Definition 5).

Since *k*-inductive invariants are a generalization of inductive invariants, we also reiterate the notion of inductive invariants as defined in our previous work.

Definition 7 (inductive invariant [5]). *Given a typed graph transformation system GTS = (R, TG) and graph constraints F and H, F is an inductive invariant for GTS under H, if, for each rule b in R, it holds that:*

$$\forall G_0, G_1((G_0 \Rightarrow_b G_1) \Rightarrow ((G_0 \vDash F \wedge G_0 \vDash H) \Rightarrow (G_1 \vDash F \vee G_1 \nvDash H)))$$

An inductive invariant (here: F) is a property that, given its validity before rule application ($G_0 \vDash F$), will also hold after rule application ($G_1 \vDash F$). In addition, we allow our system to be equipped with an additional *assumed graph constraint* (H). This constraint is assumed to be guaranteed by other means, such as additional verification steps. Typical examples include cardinality restrictions of the type graph, which, in our tool, are not automatically enforced otherwise. As defined above, rule applications involving violations of those properties are not considered as possible violations of the inductive invariant (cf. $G_0 \vDash H$, $G_1 \nvDash H$). In our approach, both types of constraints are required to be composed (assumed) forbidden patterns.

Example 8. Figure 3(a) depicts an assumed forbidden pattern H_1 of our running example implementing a cardinality constraint resulting from the physical impossibility of a shuttle being located on two tracks. Figure 3(b) (H_2) describes an undesired track topology. Both (and all other assumed forbidden patterns in our examples, see [6]) are negated and conjunctively joined in a composed assumed forbidden pattern $H = \neg H_1 \wedge \neg H_2 \wedge ... \wedge \neg H_{15}$. H is an inductive invariant and can be separately verified as such by our existing algorithm [5]. In contrast to that, the composed forbidden pattern $F = \neg F$ (Fig. 1(b)), which consists of just one forbidden pattern F, is not an inductive invariant for either variant.

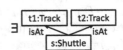

(a) Cardinality restriction

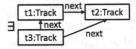

(b) Topology restriction

Fig. 3. Assumed forbidden patterns

3 k-Induction and Symbolic Encoding of Sequences

With the foundations established, we can now define the notion of k-induction [15] and k-inductive invariants for graph transformation systems. We also introduce a symbolic encoding for transformation sequences.

As established, an inductive invariant (Definition 7) is a property that, given its validity before rule application, will also hold after rule application. Likewise, a k-inductive invariant is a property whose validity in a path of length $k - 1$ ($G_z \vDash F$, below) implies its validity after the next rule application ($G_k \vDash F$).

Definition 9 (k-inductive invariant). *Given a typed graph transformation system GTS = (R, TG) and graph constraints F and H, F is a k-inductive invariant for GTS under H, if, for all sequences of transformations to R trans = $G_0 \Rightarrow_R G_1 \Rightarrow_R ... \Rightarrow_R G_k$ it holds that:*

$$\forall z (0 \leq z \leq k - 1 \Rightarrow G_z \vDash F \wedge H) \Rightarrow (G_k \vDash F \vee G_k \nvDash H)$$

As with inductive invariants, which are k-inductive invariants with $k = 1$, our formal model requires the graph constraints $\mathcal{F}$ and $\mathcal{H}$ to be a composed (assumed) forbidden pattern, respectively.

In order to deduce from the existence of a k-inductive invariant its validity in all states of an executable system, we need to consider graph grammars [7] and their initial states. In particular, a k-inductive invariant (inductive step) holds in every reachable state of the grammar if it is also valid in all transformation sequences of length $k - 1$ from the initial state (induction base).

Lemma 10. *Given a graph grammar $GG = ((\mathcal{R}, TG), G_0)$ with a k-inductive invariant $\mathcal{F}$ for $GTS = (\mathcal{R}, TG)$, $\mathcal{F}$ is satisfied in all states reachable from G_0, if the following holds:*

$$G_0 \vDash \mathcal{F} \wedge \forall i (1 \le i \le k - 1 \Rightarrow \forall G((G_0 \Rightarrow^i_{\mathcal{R}} G) \Rightarrow (G \vDash \mathcal{F})))$$

While our verification approach and contribution focus on establishing k-inductive invariants, we will shortly discuss the base of induction in our evaluation. In the following, the notion that a transformation system is *safe* (*unsafe*) will refer to the inductive step; i.e. will mean that the respective composed forbidden pattern can (cannot) be established as a k-inductive invariant.

Since there is a possibly infinite amount of transformation sequences to be analyzed in order to establish a composed forbidden pattern as a k-inductive invariant, we require a symbolic encoding for sequences of transformations. Then, reasoning over transformation sequences can be reduced to reasoning over a finite set of representative symbolic encodings. To establish such an encoding, we first require application conditions that can represent graph rule applications, similar to patterns and application conditions representing the set of graphs and morphisms that satisfy them. *Source (target) patterns* describe rule applications in an extended context (the application condition) beyond the left (right) rule side. *Target/source patterns* combine both and represent the context after one rule application and before another. Then, we combine source, target, and target/source patterns in *k-sequences of s/t (source/target) patterns* (Definition 12).

Definition 11 (source, target, and target/source pattern [5]). *A source pattern (target pattern) over a rule – specifically, over its left (right) side L (R) – is an application condition src (tar) of the form false or the form $\exists (s : L \hookrightarrow S, ac_S)$ ($\exists (t : R \hookrightarrow T, ac_T)$) with a composed negative application condition ac_S (ac_T) over S (T).*

A target/source pattern is a pair of a target and a source pattern (tar, src) which share the same codomain and application condition, i.e. $tar = \exists (t : R \hookrightarrow T, ac_T)$ and $src = \exists (s : L \hookrightarrow T, ac_T)$. A pair of morphisms (m', m) with the same codomain ($m' : R \hookrightarrow G$ and $m : L \hookrightarrow G$) satisfies a target/source pattern (tar, src), denoted $(m', m) \vDash (tar, src)$, if $m' \vDash tar$ and $m \vDash src$ by a common monomorphism $y : T \hookrightarrow G$, i.e. if there exists a monomorphism $y : T \hookrightarrow G$ such that $y \circ t = m'$, $y \circ s = m$, and $y \vDash ac_T$.

Definition 12 (k-sequences of s/t-patterns). *Given $k \geq 1$, a source pattern src_1 over a rule b_1, a target pattern tar_k over a rule b_k and a number of target/source patterns (tar_i, src_{i+1}) over a number of rules b_i $(1 \leq i \leq k-1)$, $seq = src_1 \Rightarrow_{b_1} tar_1, src_2 \Rightarrow_{b_2} ... \Rightarrow_{b_k} tar_k$ is a k-sequence of s/t-patterns.*

Satisfiability of k-sequences of s/t-patterns is defined as follows:

Given a sequence of transformations (of length k) $trans = G_0 \Rightarrow_{c_1,m_1,m'_1} ... \Rightarrow_{c_k,m_k,m'_k} G_k$ and a k-sequence of s/t-patterns $seq = src_1 \Rightarrow_{b_1} tar_1, src_2 \Rightarrow_{b_2} ... \Rightarrow_{b_k} tar_k$, trans satisfies seq, denoted as $trans \models seq$, if, for all i with $1 \leq i \leq k$, $c_i = b_i$, $m_i \models src_i$, $m'_i \models tar_i$ and, in particular, for all i with $1 \leq i \leq k-1$, $(m'_i, m_{i+1}) \models (tar_i, src_{i+1})$.

Two k-sequences of s/t-patterns seq, seq' are equivalent $(seq \equiv seq')$, if for all transformation sequences trans it holds that $trans \models seq \Leftrightarrow trans \models seq'$.

The idea of k-sequences of s/t-patterns (or simply s/t-pattern sequences) is to not only describe subsequent transformations but, with source and target patterns, additional context in which those transformations occur. As such, an s/t-pattern sequence is a symbolic encoding for the set of transformation sequences that satisfy it. The construction of specific s/t-pattern sequences for the verification of k-inductive invariants will be explained in the next section. It bears certain similarities to the notion of *E-concurrent rules* from [8].

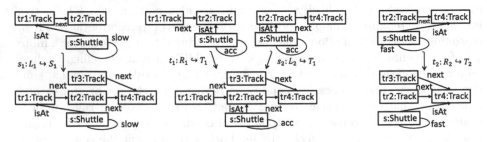

Fig. 4. Example sequence of patterns $seq = src_1 \Rightarrow_{b_1} tar_1, src_2 \Rightarrow_{b_2} tar_2$

Example 13 (s/t-pattern sequence). Figure 4 shows a 2-sequence of s/t-patterns $seq = src_1 \Rightarrow_{b_1} tar_1, src_2 \Rightarrow_{b_2} tar_2$, where rules $b_i = \langle L_i \hookleftarrow K_i \hookrightarrow R_i \rangle$ $(i = 1, 2)$ are s2a and a2f from variant 1 (Example 1). In particular, $src_1 = \exists s_1$ and $src_2 = \exists s_2$ are source patterns, $tar_1 = \exists t_1$ and $tar_2 = \exists t_2$ are target patterns, and (tar_1, src_2) is a target/source pattern. The transformation sequence *trans* (Example 3, Fig. 2) satisfies seq $(trans \models seq)$. On the other hand, no transformation sequence from variant 2 (Example 2) could satisfy seq due to the negative application condition in s2a.

As a side note, $src_1 \Rightarrow_{b_1} tar_1$ and $src_2 \Rightarrow_{b_2} tar_2$ would also be valid 1-sequences of s/t-patterns.

Since we will need to compare elements of s/t-pattern sequences to (assumed) forbidden patterns, we establish a connection between source and target patterns and (forbidden) patterns with respect to pattern implication:

Lemma 14 (reduction to pattern [5]). *Let $ac = \exists(s : L \hookrightarrow S, ac_S)$ be an application condition over L with ac_S being a composed negative application condition. For the reduction to a pattern $ac_\varnothing = \exists(i_S : \varnothing \hookrightarrow S, ac_S)$ of ac we have: For each graph G with $m : L \hookrightarrow G$ such that $m \vDash ac$, we have $G \vDash ac_\varnothing$.*

4 k-Inductive Invariant Checking

Our formal approach to verify a composed forbidden pattern as a k-inductive invariant consists of the following steps: We split the composed forbidden pattern into its individual forbidden patterns (step 1, Sect. 4.1). Then, we construct a finite set of k-sequences of s/t-patterns per forbidden pattern such that these sequences represent all transformation sequences leading to the forbidden pattern (step 2, Sect. 4.2). Finally, we analyze each s/t-pattern sequence in each set for possible violations of (assumed) forbidden patterns earlier in the sequence (step 3, Sect. 4.3). Sequences with such violations can be discarded; all others present counterexamples with respect to the validity of the k-inductive invariant.

4.1   Step 1: Separation of Forbidden Patterns

The following lemma is the formal basis for investigating individual forbidden patterns and transformation sequences that lead (see Sect. 2.1) to those patterns. It is based on the contraposition of Definition 9; its intention is to justify the procedure of finding all possible violations of individual forbidden patterns (the implication's precondition below) and trying to disprove them by finding violations earlier in the path (postcondition). The latter loosely corresponds to step 2 and 3 (Sects. 4.2 and 4.3) below.

Lemma 15. *Given a typed graph transformation system $GTS = (\mathcal{R}, TG)$ and a composed (assumed) forbidden pattern $\mathcal{F} = \bigwedge_{i \in I} \neg F_i$ ($\mathcal{H} = \bigwedge_{j \in J} \neg H_j$), $\mathcal{F}$ is a k-inductive invariant for GTS under $\mathcal{H}$, if the following holds for each k-sequence of transformations $trans = G_0 \Rightarrow_\mathcal{R} \ldots \Rightarrow_\mathcal{R} G_k$:*

$$\exists n(G_k \vDash F_n) \Rightarrow (\exists z, v(0 \le z \le k \wedge G_z \vDash H_v) \vee \exists z, v(0 \le z \le k - 1 \wedge G_z \vDash F_v))$$

4.2   Step 2: Construction of k-Sequences and Context Propagation

The following theorem describes the construction of finite sets of s/t-pattern sequences, which represent all transformation sequences leading to a specific forbidden pattern. Those sequences are possible violations of the desired k-inductive invariant.

Theorem 16 (construction of sequences). *There is a construction Seq_k such that for every pattern $F = \exists(i_P, ac_P)$, rule set $\mathcal{R}$, and k, $\mathrm{Seq}_k(\mathcal{R}, F)$ is a set of k-sequences of s/t-patterns and for each transformation sequence trans to $\mathcal{R}$ of length k that leads to F, there is a $seq \in \mathrm{Seq}_k(\mathcal{R}, F)$ with $trans \vDash seq$.*

Construction. Seq_k *is inductively constructed as follows (with appropriate indexes and index sets u, j and U, J_u, respectively), starting with Seq_1 (left figure):*

1. *For each rule $b_u = \langle (L_u \hookleftarrow K_u \hookrightarrow R_u), ac_{L_u} \rangle \in \mathcal{R}$, $\text{Shift}(i_{R_u}, F) = \vee_{j \in J_u} tar_{u,j}$ is a disjunction of target patterns over R_u of the form $tar_{u,j} = \exists (t_j, ac_{T_j})$.*
2. *For each such target pattern $tar_{u,j}$, $src'_{u,j} = \text{L}(b_u, tar_{u,j})$ is a source pattern over L_u of the form $src'_{u,j} = \text{false}$ or $src'_{u,j} = \exists (s_j, ac'_{S_j})$.*
3. *For the latter case, $src_{u,j} = \exists (s_j, ac'_{S_j} \wedge \text{Shift}(s_j, ac_{L_u}))$ is a source pattern.*
4. *For each such pair of source and target pattern $src_{u,j}$ and $tar_{u,j}$, $src_{u,j} \Rightarrow_{b_u} tar_{u,j}$ is a 1-sequence of s/t-patterns.*
5. *Finally, we define $\text{Seq}_1(\mathcal{R}, F) = \{ src_{u,j} \Rightarrow_{b_u} tar_{u,j} \mid b_u \in \mathcal{R} \wedge j \in J_u \}$ as the set of these sequences.*

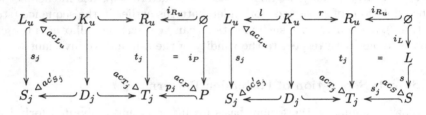

Given $\text{Seq}_k(\mathcal{R}, F)$, we construct $\text{Seq}_{k+1}(\mathcal{R}, F)$ as follows (right figure):

1. *For each sequence $seq = src_1 \Rightarrow \ldots \Rightarrow tar_k \in \text{Seq}_k(\mathcal{R}, F)$ with $src_1 = \exists (s : L \hookrightarrow S, ac_S)$ and each $b_u = \langle (L_u \hookleftarrow K_u \hookrightarrow R_u), ac_{L_u} \rangle \in \mathcal{R}$, $\text{Shift}(i_{R_u}, \exists (i_S, ac_S)) = \vee_{j \in J_u} tar_{u,j}$ is a disjunction of target patterns over R_u with $tar_{u,j} = \exists (t_j, ac_{T_j})$.*
1'. *For each such target pattern $tar_{u,j}$, $src_1^* = \exists (s'_j \circ s, ac_{T_j})$ is a source pattern.*
2. *For each such target pattern $tar_{u,j}$, $src'_{u,j} = \text{L}(b_u, tar_{u,j})$ is a source pattern over L_u of the form $src'_{u,j} = \text{false}$ or $src'_{u,j} = \exists (s_j, ac'_{S_j})$.*
3. *For the latter case, $src_{u,j} = \exists (s_j, ac'_{S_j} \wedge \text{Shift}(s_j, ac_{L_u}))$ is a source pattern.*
4. *For each such pair of source and target pattern $src_{u,j}$ and $tar_{u,j}$, $src_{u,j} \Rightarrow_{b_u} tar_{u,j}, src_1^* \Rightarrow \ldots \Rightarrow tar_k$ is a k+1-sequence of s/t-patterns.*
5. *Finally, we define $\text{Seq}_{k+1}(\mathcal{R}, F) = \{ src_{u,j} \Rightarrow_{b_u} tar_{u,j}, src_1^* \Rightarrow \ldots \Rightarrow tar_k \mid b_u \in \mathcal{R} \wedge j \in J_u \wedge seq \in \text{Seq}_k(\mathcal{R}, F) \}$ as the set of these sequences.*

Also, given a set of rules $\mathcal{R}$ and a composed forbidden pattern $\mathcal{F} = \wedge_{i \in I} \neg F_i$ with forbidden patterns F_i, we define $\text{Seq}_k(\mathcal{R}, \mathcal{F}) = \cup_{i \in I} \text{Seq}_k(\mathcal{R}, F_i)$.

To encode all situations in which a forbidden pattern is violated after a rule application, $\text{Seq}_1(\mathcal{R}, F)$ first builds target patterns for all overlappings of right rule sides and the forbidden pattern (cf. $\text{Shift}(i_{R_u}, F)$ in 1). Then, for each of those target patterns, the respective source pattern – the context before rule application – is computed ($\text{L}(b_u, tar_{u,j})$, 2). Finally, the left application condition of the applied rule is transferred from its left rule side to the context of the source pattern $(\text{Shift}(s_j, ac_{L_u}), 3)$. All pairs of a source and a target pattern thusly created then constitute a 1-sequence of s/t-patterns in $\text{Seq}_1(\mathcal{R}, F)$ (4, 5).

Given the set of sequences created by Seq_k, Seq_{k+1} repeats this process until the sequences reach a fixed length k. In particular, it creates all overlappings of right rule sides and leftmost source patterns of k-sequences $(\text{Shift}(i_{R_u}, \exists(i_S, ac_S)), 1)$ to create target patterns, then adds the newly accumulated context to the leftmost source patterns $(src_1^* = \exists(s_j' \circ s), ac_{T_j}, 1')$. 2–5 mirror the respective computations of Seq_1. Since all involved constructions (particularly Shift and L) produce finite results, Seq_k always yields finite results for a fixed k.

Theorem 16 states that all transformation sequences of length k that lead to a forbidden pattern F have a representative s/t-pattern sequence in $\text{Seq}_k(\mathcal{R}, F)$. We also proof that each s/t-pattern sequence is meaningful in the sense that every transformation sequence it represents actually leads to the forbidden pattern:

Lemma 17. *Given a set of rules* $\mathcal{R}$, *a pattern* F, *and the set* $\text{Seq}_k(\mathcal{R}, F)$, *for each k-sequence of s/t-patterns seq* $\in \text{Seq}_k(\mathcal{R}, F)$, *every transformation sequence trans with trans* $\models$ *seq leads to* F.

Example 18. **Figure 4** in Example 13 also serves as an example of one sequence (of many) found in $\text{Seq}_2(\mathcal{R}, F)$ – F as in Fig. 1(b) – for variant 1 of our example. However, the context of the second rule application described by the target pattern tar_2 does not yet take accumulated context of subsequent Seq_k-constructions into account; in particular, it lacks the fourth track (tr1) required in transformation sequences satisfying *seq*. Also, note that our transformation sequence *trans* with *trans* $\models$ *seq* (Example 13) leads to F (cf. Lemma 17).

For variant 2 of our example system, the Seq_2-construction for the case above would calculate $\text{Shift}(s_1, ac_L)$, where ac_L is the additional negative application condition of graph rule s2a (Fig. 1(f)). Since ac_L – forbidding the existence of a subsequent switch – actually exists in S_1, the result of $\text{Shift}(s_1, ac_L)$ would (upon evaluation) default to false. Since no transformation sequence can satisfy such a sequence of s/t-patterns, the sequence is invalid as a counterexample, which would become apparent in our analysis in step 3 (Sect. 4.3).

Repeating the Seq_k-construction only accumulates context in backward direction by reverse rule applications (via L). Similarly, acquired context can also be propagated in forward direction. In particular, our construction below uses L to recursively propagate context from the leftmost source pattern over the respective rules through the whole sequence. To justify the process, Lemma 19 establishes that the set of all transformation sequences represented by the s/t-pattern sequences in Seq_k equals the set of transformation sequences represented by the propagated s/t-pattern sequences in Seq_k. Although that set of transformation sequences remains unchanged, *forward propagation* enriches our symbolic representation in order to discard false negatives in the subsequent analysis step.

Lemma 19 (forward propagation over sequences). *Given a set of graph rules $\mathcal{R}$, a pattern F, and the set of sequences constructed by $\mathrm{Seq}_k(\mathcal{R}, F)$, we describe forward propagation as a function* prop *such that for all* $seq \in \mathrm{Seq}_k(\mathcal{R}, F)$, *we have* $seq \equiv \mathrm{prop}(seq)$.

Construction. *We construct* prop *recursively as follows:*

$$\mathrm{prop}(src_1 \Rightarrow_{b_1} tar_1) = src_1 \Rightarrow_{b_1} tar'_1$$

$$\mathrm{prop}(src_1 \Rightarrow_{b_1} tar_1, \qquad\qquad src_2 \Rightarrow_{b_2} tar_2, ..., src_k \Rightarrow_{b_k} tar_k)$$
$$= src_1 \Rightarrow_{b_1} tar'_1, \mathrm{prop}(\mathrm{comb}(tar'_1, src_2) \Rightarrow_{b_2} tar_2, ..., src_k \Rightarrow_{b_k} tar_k),$$

where $tar'_1 = \mathrm{L}(b_1^{-1}, src_1)$ *and*

$$\mathrm{comb}(\mathit{false}, src_2) = \mathit{false}$$
$$\mathrm{comb}(tar'_1, \mathit{false}) = \mathit{false}$$
$$\mathrm{comb}(tar'_1, src_2) = \exists(t'_1 \circ s'_2 \circ s_2, ac_{T'_1})$$

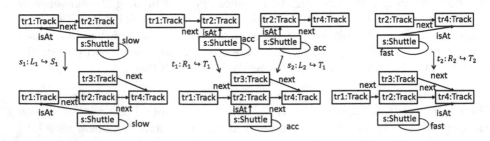

as in the diagram on the right, with $src_1 = \exists(t'_0 \circ s_1, ac_{T'_0})$, $tar_1 = \exists(t_1, ac_{T_1})$, $tar'_1 = \exists(t'_1 \circ t_1, ac_{T'_1}) = \mathrm{L}(b_1^{-1}, src_1)$, *and* $src_2 = \exists(s'_2 \circ s_2, ac_S)$.

Note that for the first call of prop *on a sequence constructed by* Seq_k, *we have* $src_1 = \exists(s_1, ac_{S_1})$ *and* T'_0 *and* T'_1 *will not exist. For the purpose of* prop *and* comb, T'_0 *and* S_1 *can be treated as isomorphic (with* $ac_{T'_0} = ac_{S_1}$); *then,* T_1 *and* T'_1 *are isomorphic as well,* t'_1 *is an isomorphism, and* $ac_{T_1} = ac_{T'_1}$.

![Sequence diagram figures]

Fig. 5. Sequence $seq' = \mathrm{prop}(seq) = src_1 \Rightarrow_{b_1} tar_1, src_2 \Rightarrow_{b_1} tar'_2$, with $seq \equiv seq'$

Example 20. Figure 5 shows the 2-sequence of s/t-patterns $seq' = \mathrm{prop}(seq)$. The difference to seq (Fig. 4, Example 13) lies in the additional context (tr1) in target pattern tar_2. This also exemplifies the intention of calculating prop: additional information may make a difference in our subsequent analysis (step 3, below).

4.3   Step 3: Analysis of Sequences

Our final and central theorem describes the main result of our approach and formalization: the analysis of s/t-pattern sequences created by our earlier steps.

Theorem 21 (k-inductive invariant checking). *Let* $GTS = (\mathcal{R}, TG)$ *be a graph transformation system and* $\mathcal{F} = \bigwedge_{i \in I} \neg F_i$ *(*$\mathcal{H} = \bigwedge_{j \in J} \neg H_j$*) be composed (assumed) forbidden patterns. Let* $\mathrm{Seq}_k(\mathcal{R}, F)$ *be the set of k-sequences constructed from the pattern* F *and* $\mathrm{Seq}_k(\mathcal{R}, \mathcal{F}) = \cup_{i \in I} \mathrm{Seq}_k(\mathcal{R}, F_i)$. *Let, for a source (target) pattern* src_z *(*tar_z*),* $src_{z,\varnothing}$ *(*$tar_{z,\varnothing}$*) be the reduction of* src_z *(*tar_z*) to a pattern.*

$\mathcal{F}$ *is a k-inductive invariant under* $\mathcal{H}$ *if, for all sequences* $\mathrm{prop}(seq) = src_1 \Rightarrow_{b_1}$ $\dots \Rightarrow_{b_k} tar_k$ *with* $seq \in Seq_k(\mathcal{R}, \mathcal{F})$, *one of the following conditions holds:*

1. $\exists z, v (1 \le z \le k \wedge (src_{z,\varnothing} \vDash H_v \vee src_{z,\varnothing} \vDash F_v))$.   2. $\exists v (tar_{k,\varnothing} \vDash H_v)$.

Example 22. For $k = 2$, there is a counterexample for variant 1 (Fig. 5, Example 20). All sequences of length 2 for variant 2 would be discarded by Theorem 21. Hence, $\neg F$ is a 2-inductive invariant for variant 2, but not for variant 1.

Our approach is sound in the sense that for every violating transformation sequence, a symbolic counterexample (s/t-pattern sequence) will be found. It is not necessarily complete: spurious counterexamples can occur, because the theorem above only describes a sufficient condition. Addressing this approach-inherent drawback requires a more complex notion of pattern implication, which is, in general, an undecidable problem [13]. However, previous [5] and current evaluation show the applicability of our approach even without such extensions.

Our implementation and tool closely follow the formalization established above. Given a fixed value for k, a set of graph rules, and a composed forbidden pattern to be verified as a k-inductive invariant under a composed assumed forbidden pattern, the tool: (a) constructs all 1-sequences of s/t-patterns leading to forbidden patterns (Theorem 16), (b) analyzes those sequences (Theorem 21), (c) prolonges the remaining sequences by one step (Theorem 16), (d) applies forward propagation (Lemma 19) and (e) analysis (Theorem 21, again), and then repeats (c), (d), and (e) until the sequences' length reaches k. If all sequences of length k have been discarded, the composed forbidden pattern is a k-inductive invariant. Otherwise, the remaining sequences serve as counterexamples. Because of the finiteness of all involved constructions, this algorithm always terminates.

5   Evaluation

In the following, we discuss the experimental evaluation of our approach, which we implemented as an extension of our tool described in [1,5]. We considered variants 1 (Example 1) and 2 (Example 2) of our running example as two cases where a k-inductive invariant cannot and can (for $k = 2$) be established, i.e., as cases for an unsafe and a safe system. Variants 3 and 4 (see [6]) present two more elaborate cases, which include sensor faults and a *single fault assumption*.

First, we used our existing tool for the verification of (1-)inductive invariants [5] for all example variants ($k = 1$). Then, we used our extensions implementing the algorithm formalized in this paper ($k > 1$). We consider configurations with and without forward propagation[1] (Lemma 19), configurations that compute all counterexamples (denoted by *full*), and configurations that enforce termination as soon as one counterexample of length k has been found (*stop on ce*).

Our results[2] are shown in Table 1. The numbers in brackets denote the number of rules, forbidden properties, and assumed forbidden properties for the respective variants. Columns k, c, and t denote the length of the path for the inductive step, the number of counterexamples, and runtime in seconds, respectively. Column r shows the verification result, which can take the values false (f) for an unsafe system, fn for false negatives (spurious counterexamples), $f+fn$ for a combination of both, true (t) for a safe system, or na (not applicable).

The term *false negative* refers to counterexamples, i.e. s/t-pattern sequences of the respective length k, for which there cannot exist a satisfying transformation sequence that describes an actual violation of the k-inductive invariant. Such counterexamples may occur (1) if forward propagation is not considered, which leads to incomplete information during the analysis, and (2) if a more complex (potentially undecidable) notion of pattern implication is required (cf. Section 4.3). However, since all forbidden and assumed forbidden patterns in our examples are of the simple form $F = \exists(i_P : \varnothing \hookrightarrow P, \text{true})$, the second type of false negatives cannot occur here. Hence, all counterexamples resulting from experiments that include forward propagation are true negatives (f) and could be instantiated as transformation sequences that violate the k-inductive invariant.

For the *full* case with forward propagation, the property is not a k-inductive invariant (up to $k = 6$) for the erroneous variant 1, as expected (cf. Examples 1 and 22). The corrections resulting in variant 2 (Example 2) lead to a safe system where the property can be established as a 2-inductive invariant. Likewise, variant 4 is a fixed version of the erroneous variant 3. While the computational effort for variants 2 and 4 is minimal, variants 1 and 3 show strongly increasing numbers for counterexamples and computation time. This will almost always be the case for erroneous systems. However, in both cases, execution with the *stop on ce* option will quickly return results for manual inspection of unsafe systems.

As explained before, execution without forward propagation can lead to false negatives. In particular, without forward propagation the safety property cannot be established even for the (actually safe) variant 4. Also, the high number of false negatives in variant 3 leads to even higher numbers of (false) counterexamples for subsequent values of k and significantly higher computation times.

---

[1] To allow verification without forward propagation, Theorem 21 can be modified by considering all $seq \in \text{Seq}_k(\mathcal{R}, \mathcal{F})$ instead of all $\text{prop}(seq) \in \text{Seq}_k(\mathcal{R}, \mathcal{F})$.

[2] Setup: 64-bit system, two cores at 2.8 GHz, 8 GB main memory, Eclipse 4.5.1, Java 8, Windows 7. Java heap space limit was set to 1 GB, with the exception of variant 4 with forward propagation and $k = 6$, which required 4 GB.

Table 1. Results for 1- and *k*-inductive invariant checking ([5] and current approach).

example (# rules/ forbidden/ assumed)	k	without forward propagation						with forward propagation					
		full			stop on ce			full			stop on ce		
	k	c	t	r	c*	t	r	c	t	r	c*	t	r
variant 1 (7/1/15), unsafe	1	6	<1	f	1	<1	f			na			na
	2	9	<1	f	1	<1	f	9	<1	f	1	<1	f
	3	47	1.0	f	1	<1	f	47	1.0	f	1	<1	f
	4	217	2.8	f	1	<1	f	217	2.9	f	1	<1	f
	5	1102	15.4	f+fn	1	<1	f/fn	1063	15.6	f	1	<1	f
	6	6211	95.7	f+fn	1	<1	f/fn	5551	94.1	f	1	<1	f
variant 2 (7/1/15), safe	1	6	<1	f	1	<1	f			na			na
	2	0	<1	t	0	<1	t	0	<1	t	0	<1	t
variant 3 (10/1/16), unsafe	1	9	<1	f	1	<1	f			na			na
	2	15	<1	f+fn	1	<1	f/fn	6	<1	f	1	<1	f
	3	128	1.5	f+fn	1	<1	f/fn	27	1.1	f	1	<1	f
	4	737	10.3	f+fn	1	<1	f/fn	100	2.7	f	1	<1	f
	5	4389	78.7	f+fn	1	<1	f/fn	444	12.1	f	1	<1	f
	6	28514	741.0	f+fn	1	<1	f/fn	2011	74.2	f	1	1	f
variant 4 (10/1/16), safe	1	6	<1	f	1	<1	f			na			na
	2	9	<1	fn	1	<1	fn	0	<1	t	0	<1	t

To establish the base of induction (see Lemma 10) for a graph grammar with an initial graph, we used the model checker GROOVE [9] to (successfully) check all paths of length 1 ($k - 1 = 2 - 1$) from the initial graph for variants 2 and 4.

An important issue is the choice of k for the verification. There exist cases (variants 2 and 4) where the desired property can be established as a k-inductive invariant for small k, but not as a 1-inductive invariant. If the estimated value of k for an invariant is not known, we suggest to verify systems with increasing k, starting with 1, and to use counterexamples to fix system errors, as seen in variants 1 (fixed in variant 2) and 3 (fixed in 4). While we are confident that the technique is also applicable for different examples of similar size and values for k, we cannot yet generalize that claim for larger examples. However, since the approach's complexity is independent from a system's state space, it may be applied where approaches based on the state space are impractical.

6 Conclusion and Outlook

We presented an approach for automatic verification of k-inductive invariants that supports reasonably expressive graph rules and properties. We have proven and implemented our approach, which employs a finite symbolic encoding of traces. Further, our evaluation has demonstrated that k-inductive invariants can be established for some examples where inductive invariants are not sufficient.

Moving on, we plan to study further options to enrich k-sequences, optimize our algorithm, and apply suitable counterexample-guided refinement techniques.

Acknowledgments. We would like to thank Leen Lambers for her comprehensive feedback on a draft version of this paper.

References

1. Becker, B., Beyer, D., Giese, H., Klein, F., Schilling, D.: Symbolic invariant verification for systems with dynamic structural adaptation. In: Proceedings of the 28th International Conference on Software Engineering (ICSE). ACM, New York (2006)
2. Blume, C., Bruggink, H.J.S., Engelke, D., König, B.: Efficient symbolic implementation of graph automata with applications to invariant checking. In: Ehrig, H., Engels, G., Kreowski, H.-J., Rozenberg, G. (eds.) ICGT 2012. LNCS, vol. 7562, pp. 264–278. Springer, Heidelberg (2012). doi:10.1007/978-3-642-33654-6_18
3. Boneva, I.B., Kreiker, J., Kurban, M.E., Rensink, A., Zambon, E.: Graph abstraction and abstract graph transformations (amended version). Technical report TR-CTIT-12-26, University of Twente, Enschede (2012)
4. Donaldson, A.F., Haller, L., Kroening, D., Rümmer, P.: Software verification using k-induction. In: Yahav, E. (ed.) SAS 2011. LNCS, vol. 6887, pp. 351–368. Springer, Heidelberg (2011). doi:10.1007/978-3-642-23702-7_26
5. Dyck, J., Giese, H.: Inductive invariant checking with partial negative application conditions. In: Parisi-Presicce, F., Westfechtel, B. (eds.) ICGT 2015. LNCS, vol. 9151, pp. 237–253. Springer, Cham (2015). doi:10.1007/978-3-319-21145-9_15
6. Dyck, J., Giese, H.: k-Inductive Invariant Checking for Graph Transformation Systems. Technical report, University of Potsdam (2017)
7. Ehrig, H., Ehrig, K., Prange, U., Taentzer, G.: Fundamentals of Algebraic Graph Transformation. Springer, Secaucus (2006)
8. Ehrig, H., Golas, U., Habel, A., Lambers, L., Orejas, F.: M-adhesive transformation systems with nested application conditions. part 1: parallelism, concurrency and amalgamation. Math. Struct. Comput. Sci. 24, 1–48 (2014)
9. Ghamarian, A.H., de Mol, M.J., Rensink, A., Zambon, E., Zimakova, M.V.: Modelling and analysis using GROOVE. Int. J. Softw. Tools Technol. Transf. 14(1), 15–40 (2012)
10. Habel, A., Pennemann, K.-H.: Correctness of high-level transformation systems relative to nested conditions. Math. Struct. Comput. Sci. 19, 1–52 (2009)
11. König, B., Kozioura, V.: Augur 2 - a new version of a tool for the analysis of graph transformation systems. Electron. Notes Theoret. Comput. Sci. 211, 201–210 (2008)
12. König, B., Stückrath, J.: A general framework for well-structured graph transformation systems. In: Baldan, P., Gorla, D. (eds.) CONCUR 2014. LNCS, vol. 8704, pp. 467–481. Springer, Heidelberg (2014). doi:10.1007/978-3-662-44584-6_32
13. Pennemann, K.-H.: Development of correct graph transformation systems. Ph.D. thesis, University of Oldenburg (2009)
14. Schmidt, Á., Varró, D.: CheckVML: a tool for model checking visual modeling languages. In: Stevens, P., Whittle, J., Booch, G. (eds.) UML 2003. LNCS, vol. 2863, pp. 92–95. Springer, Heidelberg (2003). doi:10.1007/978-3-540-45221-8_8
15. Sheeran, M., Singh, S., Stålmarck, G.: Checking safety properties using induction and a SAT-solver. In: Hunt, W.A., Johnson, S.D. (eds.) FMCAD 2000. LNCS, vol. 1954, pp. 127–144. Springer, Heidelberg (2000). doi:10.1007/3-540-40922-X_8
16. Steenken, D.: Verification of infinite-state graph transformation systems via abstraction. Ph.D. thesis, University of Paderborn (2015)

Probabilistic Timed Graph Transformation Systems

Maria Maximova[1](✉), Holger Giese[1], and Christian Krause[2]

[1] Hasso Plattner Institute at the University of Potsdam, Potsdam, Germany
{maria.maximova,holger.giese}@hpi.de
[2] SAP SE, Potsdam, Germany
christian.krause01@sap.com

Abstract. Today, software has become an intrinsic part of complex distributed embedded real-time systems. The next generation of embedded real-time systems will interconnect the today unconnected systems via complex software parts and the service-oriented paradigm. Therefore besides timed behavior and probabilistic behavior also structure dynamics, where the architecture can be subject to changes at run-time, e.g. when dynamic binding of service end-points is employed or complex collaborations are established dynamically, is required. However, a modeling and analysis approach that combines all these necessary aspects does not exist so far.

To fill the identified gap, we propose Probabilistic Timed Graph Transformation Systems (PTGTSs) as a high-level description language that supports all the necessary aspects of structure dynamics, timed behavior, and probabilistic behavior. We introduce the formal model of PTGTSs in this paper and present a mapping of models with finite state spaces to probabilistic timed automata (PTA) that allows to use the PRISM model checker to analyze PTGTS models with respect to PTCTL properties.

1 Introduction

Today, software has become an intrinsic part of complex distributed embedded real-time systems, which need to realize more advanced functionality. The next generation of embedded real-time systems will interconnect the today unconnected systems via complex software parts and the service-oriented paradigm. It is envisioned that such networked systems will be able to behave much more intelligently by building communities of autonomous agents that exploit local and global networking to adapt and optimize their functionality [6].

In contrast to today's real-time systems, their behavior will in addition be characterized by *structure dynamics* that results from their complex coordination behavior. This structure dynamics requires execution in real-time and reconfiguration at run-time to adjust the systems behavior to its changing context and goals, leading to self-adaptation and self-optimization [25]. For these systems, also the structure resp. architecture is subject to changes at run-time, e.g. when

© Springer International Publishing AG 2017
J. de Lara and D. Plump (Eds.): ICGT 2017, LNCS 10373, pp. 159–175, 2017.
DOI: 10.1007/978-3-319-61470-0_10

dynamic binding of service end-points is employed or complex collaborations are established dynamically. In the latter case, often the structural context in the form of local topology and distribution information is particularly important.

As a concrete example for such an advanced embedded real-time system, the RailCab research project [24] aims at combining a passive track system with intelligent shuttles that operate autonomously, act individually, and make independent and decentralized operational decisions. For the RailCab application example it holds that some functionality may be safety-critical such as the convoy coordination, or mission-critical for economic reasons such as the negotiation of the transport contracts. Furthermore, the required properties are not merely qualitative ones but also quantitative ones involving time as well as probabilities. For instance, convoy coordination protocols have to be established between shuttles nearby in the topology, usually involving hard real-time constraints, and the sent protocol message may be lost with a non-zero probability. Consequently, we need methods and tools to guarantee critical quantitative properties when developing such systems, which include *structure dynamics*, *timed behavior*, and *probabilistic behavior*.

Combinations of different modeling approaches have led to a number of new interesting applications in the last couple of years. In the following, we briefly describe related modeling and analysis approaches, which combine some of the aspects of *structure dynamics*, *timed behavior*, and *probabilistic behavior*.

Timed graph transformation systems (TGTSs) [4,10,22] facilitate the modeling of timed behavior in graph transformation systems using timed automata concepts.[1] Specifically, nodes can be annotated with real-valued clocks which can be dynamically added and removed from the systems. Rules can include clock constraints as additional application conditions, and clocks can be reset. Using symbolic, *zone*-based representations [7,18] and an implementation in an extension [22] of the GROOVE tool [15], the state spaces of TGTSs can be explored and analyzed, e.g. for time-bounded reachability checks. Moreover, inductive invariant checking [4] for TGTSs provides a means to deal with infinite-state systems. Thus, TGTSs enable the analysis of combined models with structure dynamics and real-time behavior. However, probabilistic behavior is not supported.

A combination of structure dynamics and probabilistic behavior is supported by probabilistic graph transformation systems (PGTSs) [16], which are an extension of the graph transformation theory with discrete probabilistic behavior. In PGTSs, transformation rules are allowed to have multiple right-hand sides, where each of them is annotated with a probability. The choice for a rule match is nondeterministic, whereas the effect of a rule is probabilistic. This approach can be used to model randomized behavior and on-demand probabilistic failures, such as message loss in unreliable communication channels and supports modeling

---

[1] An alternative approach for graph transformation systems with time was developed in [11]. However, this approach is not suitable in our context since symbolic state space representations and quantitative analysis methods are not considered in [11].

and analysis by an extension of the HENSHIN [8] tool and a mapping to the PRISM [17] model checker.[2]

Real-time rewrite theories as supported by the executable specification language of Real-time MAUDE [23] facilitate combined modeling of structure dynamics and real-time behavior. Analysis goals include reachability checks for failures of safety properties and model checking of time-bounded temporal logic properties. Such properties are in general not decidable and therefore the provided tool support is incomplete.

Probabilistic rewrite theories implemented in PMAUDE [1] provide a combination of structure dynamics, probabilistic behavior for discrete branching, and stochastic timed behavior. Properties for PRTs are specified using probabilistic temporal logic and checked using discrete event simulation, e.g. using the VESTA tool [27]. However, in order to simulate and analyze models in PMAUDE, *all* nondeterminism has to be resolved, i.e., neither discrete nondeterministic choice nor timed nondeterminism as required for real-time behavior, are allowed.

Probabilistic Timed Automata (PTA) [19] combine the modeling features of Markov decision processes (MDPs) [5] and timed automata (TA) [2,3] and thereby allow to analyze systems exhibiting both timed and probabilistic phenomena. Analysis goals for PTA include the checking of probabilistic time-bounded reachability, computation of rewards, as well as PTCTL model checking [19]. Such properties can be analyzed for PTA, e.g., using the PRISM tool.

The timed and probabilistic extensions of rewrite systems, specifically rewrite theories in MAUDE variants and GTSs, provide the best coverage for the required modeling features. However, none of the existing models facilitates the modeling and analysis of *all* identified requirements.

To fill the identified gap, we propose to combine and extend the existing models to the formalism of Probabilistic Timed Graph Transformation Systems (PTGTSs) that supports modeling and analysis of *structure dynamics, timed behavior*, and *probabilistic behavior*. We introduce the formal model of PTGTSs in this paper and present a mapping of models with finite state spaces to probabilistic timed automata (PTA) that allows to use the PRISM model checker to analyze PTGTS models with respect to PTCTL properties.

The paper is structured as follows. First, the necessary prerequisites in form of probabilistic timed automata (PTA) are recapitulated in Sect. 2. Then, we introduce Probabilistic Timed Graph Transformation Systems (PTGTSs) in Sect. 3. Subsequently in Sect. 4, we present the tool support for our approach using the graph transformation tool HENSHIN and apply it to model our running example handling a shuttle scenario. Finally in Sect. 5, we consider the analysis of PTGTS models by combining the state space generation of HENSHIN and the

---

[2] Also stochastic graph transformation systems (SGTSs) [13] that incorporate stochastic timed behavior into GTSs by including continuous-time probability distributions that describe the average delay of firing of rules, once they are enabled, have been proposed. However, note that they do neither support probabilistic behavior nor real-time behavior as they assume a different model of time.

PTA model checking of the PRISM tool via a mapping. The paper is closed with some final conclusions and an outlook on planned future work.

2   Probabilistic Timed Automata

In this section we first informally introduce the formalisms of probabilistic [26] resp. timed automata [2,3] and then combine them to the notion of probabilistic timed automata [19] used for modeling of real-time systems with probability. Probabilistic automata (PA) were introduced in [26] to add probabilistic choice to finite automata by assigning to each edge a probability. The notion of *discrete probability distribution* plays a central role in the context of PA. For a denumerable set A a discrete probability distribution is given by the function $\mu : A \rightarrow [0,1]$ with $\sum_{a \in A} \mu(a) = 1$. Furthermore, $Dist(A)$ denotes the set of all discrete probability distributions $\mu : A \rightarrow [0,1]$.

Timed automata (TA) [2,3] have proven to be a very successful modeling and analysis formalism for real-time systems such as embedded software. TA extend finite automata by making use of clocks, which restrict the behavior of the TA based on invariants, guards, and clock resets making use of constraints over clocks. For a set X of clocks $\Phi(X)$ denotes the set of all clock constraints ϕ generated by $\phi ::= x_i \sim c \mid x_i - x_j \sim c \mid \phi \wedge \phi$ where $\sim \in \{<,>,\leq,\geq\}$, $c \in \mathbb{N} \cup \{\infty\}$ are constants, and $x_i, x_j \in X$ are clocks.

The configurations of TA consist of the current location of the automaton and an assignment of each clock to a current clock value given as a real number, called clock valuation. This clock valuation is used to evaluate clock constraints introduced before to restrict the behavior of an automaton.

Definition 1 (Clock Valuation). *For a set X of clocks $\mathcal{V}(X)$ denotes the set of all functions $v : X \rightarrow \mathbb{R}$ called* clock valuations, *which are also used in the context of the following notions:*

- *Clock Reset: Let $v : X \rightarrow \mathbb{R}$ and $X' \subseteq X$. Then $v[X' := 0] : X \rightarrow \mathbb{R}$ is a clock reset such that for any $x \in X$ holds if $x \in X'$ then $v[X' := 0](x) = 0$ else $v[X' := 0](x) = v(x)$.*
- *Clock Increment: Let $v : X \rightarrow \mathbb{R}$ and $\delta \in \mathbb{R}$. Then $v + \delta : X \rightarrow \mathbb{R}$ is a clock increment such that for any $x \in X$ holds $(v + \delta)(x) = v(x) + \delta$.*
- *Clock Constraint Satisfaction: Let $v : X \rightarrow \mathbb{R}$ and ϕ be some constraint over X. Then $v \models \phi$ denotes that v satisfies the constraint ϕ.*
- *Initial Clock Valuation: $v_0 : X \rightarrow \mathbb{R}$ is the initial clock valuation if $v_0(x) = 0$ for every $x \in X$. $\mathcal{V}_0(X)$ is the singleton set containing the (unique) initial clock valuation.*

The formalism of probabilistic timed automata (PTA) is an extension of TA. PTA allow for nondeterministic system behavior and, in addition, a probabilistic choice between follower states using discrete probability distributions over edges. An important feature of PTA are invariants given by clock constraints. Invariants enable the specification of upper time bounds for steps to be executed and, hence, restrict the set of admissible reachable states of a system. In the following we consider the formal definition of PTA in the sense as introduced in [19].

Definition 2 (Probabilistic Timed Automata). *A tuple* $A = (S, L_{AP}, s_0,$ $X, I, P, \tau)$ *is a* probabilistic timed automaton (PTA) *if*

- S *is a finite set of locations,*
- $L_{AP} : S \to 2^{AP}$ *is a labeling function assigning to each location the set of atomic propositions that are true in that location,*
- s_0 *is an initial location with* $s_0 \in S$,
- X *is a finite set of clocks,*
- $I : S \to \Phi(X)$ *is a function assigning to each location a clock constraint (also called an invariant),*
- $P : S \to 2_{\text{fn}}^{Dist(S \times 2^X)}$ *is a function assigning to each location a finite non-empty set[3] of discrete probability distributions containing follower locations and corresponding clock resets,*
- $\tau = (\tau_s)_{s \in S}$ *is a family of functions where, for any* $s \in S$, $\tau_s : P(s) \to \Phi(X)$ *assigns to each* $p \in P(s)$ *a clock constraint (also called a guard).*

The single step relation describes the behavior of PTA by defining two kinds of steps: timed steps where all clock values are increased by the time elapsed and transition steps where a PTA switches states when allowed by the current clock values according to the used probability distributions without elapsing of time.

Definition 3 (Single Step Relation). *Let* $A = (S, L_{AP}, S_0, X, I, P, \tau)$ *be a PTA and states of A elements of* $S \times \mathcal{V}(X)$. *Then the* single step relation *is given as follows:*

- **Timed Step:** $(s, v) \xrightarrow{\delta} {}_{A}^{\text{PTA}}(s, v + \delta)$ *if* $\delta > 0$ *and for each* δ' *it holds that* $0 \le \delta' \le \delta$ *implies that* $v + \delta' \models I(s)$,
- **Transition Step:** $(s, v) \xrightarrow{\mu} {}_{A}^{\text{PTA}}(s', v[X' := 0])$ *if* $X' \subseteq X$, $\mu \in P(s)$, $\mu(s', X') > 0$, $v \models \tau_s(\mu)$, *and* $v[X' := 0] \models I(s')$.

For a concrete example of a PTA and its step relation see [21, Chap. 2].

According to [19], the underlying model for PTA is given by so-called probabilistic timed structures (PTSs). A PTS is a variant of a Markov decision process (MDP) [5], which is obtained by extension of a timed structure [14] with the probabilistic choice over transitions, i.e., the transition function *Steps* of a PTS results in a choice over pairs consisting of a duration of a transition and a discrete probability distribution over the follower states.

Definition 4 (Probabilistic Timed Structure). *A probabilistic timed structure (PTS)* $\mathcal{M} = (Q, Steps, L_{AP})$ *is a labeled MDP where*

- Q *is a set of states,*
- $Steps : Q \to 2^{\mathbb{R} \times Dist(Q)}$ *is a transition function assigning to each state* $q \in Q$ *a set* $Steps(q)$ *of pairs* (t, p), *where* $t \in \mathbb{R}$ *is a duration of a transition and* $p \in Dist(Q)$ *is a discrete probability distribution over the follower states,*
- $L_{AP} : Q \to 2^{AP}$ *is a state labeling function.*

---

[3] For an arbitrary set M, 2_{fn}^{M} denotes the set of finite nonempty subsets of M.

Besides the definition of PTA behavior in the form of the single step relation, we also define in the following how PTA give rise to PTSs to enable the comparison of PTA and PTGTS models later on and to be able to make use of the PTCTL logic [19], which has a semantics defined on PTSs as well.

Definition 5 (Induced PTS for a PTA). Let $A = (S, L_{AP}, s_0, X, I, P, \tau)$ be a PTA. Then $\mathcal{M}_A = (Q_A, Steps_A, L_A)$ is the induced PTS if

- $Q_A = \{(s, v) \mid (s_0, v_0) \xrightarrow{*}_A \mathrm{PTA}(s, v)\}$,
- $L_A(s, v) = L_{AP}(s)$,
- $Steps_A(s, v) = \{(\delta, \nu) \mid (s, v) \xrightarrow{\delta}_A \mathrm{PTA}(s, v + \delta) \ \wedge \ \nu(s, v + \delta) = 1\}$
  $\cup \{ (0, \nu) \mid (s, v) \xrightarrow{\mu}_A \mathrm{PTA}(s', v[X' := 0])$
  $\wedge \ \nu(\overline{s}, \overline{v}) = \sum_{\mu \in P(s):\ \mu(s'', X'') > 0\ \wedge\ (s,v) \xrightarrow{\mu}_A \mathrm{PTA}(s'', v[X'' := 0]) = (\overline{s}, \overline{v})} \mu(s'', X'')\}.$

For the case of a transition step we define the probability distribution ν on the possible follower states such that if alternative PTA steps result in identical PTA configurations, the probabilities for their occurrence are added, as required to obtain a well-defined probability distribution.

We use the PTCTL logic defined on PTSs to state various relevant properties on probabilistic real-time systems (see [21, Chap. 2] for detailed definitions). In our running example, introduced in Sect. 4, we state the desired probabilistic reachability properties and verify them using the PRISM model checker, which is able to check a subset of PTCTL using multiple back-end engines.

3   Probabilistic Timed Graph Transformation Systems

In this section we recall the framework of graph transformation systems (GTSs) and introduce a new formalism of Probabilistic Timed Graph Transformation Systems (PTGTSs) allowing for modeling and analysis of structure dynamics, timed behavior as well as probabilistic behavior of systems.

In context of our approach we focus on the formalism of typed graphs. A *graph* $G = (G_V, G_E, s_G, t_G)$ consists of a set G_V of nodes, a set G_E of edges, and source and target functions $s_G, t_G : G_E \to G_V$. For two given graphs $G = (G_V, G_E, s_G, t_G)$ and $H = (H_V, H_E, s_H, t_H)$, a *graph morphism* $f : G \to H$ is a pair of mappings $f_V : G_V \to H_V$, $f_E : G_E \to H_E$ compatible with the source and target functions, i.e., $f_V \circ s_G = s_H \circ f_E$ and $f_V \circ t_G = t_H \circ f_E$.

Let TG be a distinguished graph, called a type graph. Then a typed graph is given by a tuple $(G, type)$ consisting of a graph G together with a graph morphism $type : G \to TG$. For two given typed graphs $G'_1 = (G_1, type_1)$ and $G'_2 = (G_2, type_2)$, a *typed graph morphism* $f : G'_1 \to G'_2$ is a graph morphism $f : G_1 \to G_2$ compatible with the typing functions, i.e., $type_2 \circ f = type_1$.

The adaptation of graphs can be realized using graph transformation rules, which are to be understood as local rule-based modifications defining additions and removals of substructures. A rule $\rho = L \xleftarrow{l} K \xrightarrow{r} R$ is given by a span of injective typed graph morphisms with the graphs L and R called the left-hand resp. the right-hand side of the rule. The transformation procedure defining a graph transformation step is formally introduced by the *DPO approach* [9].

In the following we introduce the new formalism of PTGTSs. We assume here that all graphs considered in the context of PTGTSs are typed over some type graph TG containing at least a type node *Clock*. Furthermore, for every graph G we use the function $CN(G) = \{n \mid n \in G_V \land type_V(n) = Clock\}$ returning all nodes of the type *Clock* contained in G to identify in every graph the nodes used for time measurement only. In the following we call such identified nodes simply *clocks*.

The formalism of PTGTSs is a combination of Probabilistic Graph Transformation Systems (PGTSs) [16] and Timed Graph Transformation Systems (TGTSs) [4,10,22]. Similarly to PGTSs, transformation rules in PTGTSs can have multiple right-hand sides, where each of them is annotated with a probability. The choice for a rule match is nondeterministic, whereas the effect of a rule is probabilistic. Similarly to TGTSs, each probabilistic timed graph transformation rule has a guard formulated over clocks contained in the left-hand side of the rule, which is used to control the rule application. Moreover, each rule contains the information about clocks that have to be reset during the rule application.

Definition 6 (Probabilistic Timed Graph Transformation Rule). $R = (L, P, \mu, \phi, r_C)$ *is a probabilistic timed graph transformation rule if*

- L *is a common left-hand side graph,*
- P *is a finite set of graph transformation rules ρ with $lhs(\rho) = L$, where $lhs(\rho)$ provides the left-hand side of the rule ρ,*
- $\mu \in Dist(P)$ *is a probability distribution,*
- $\phi \in \Phi(CN(L))$ *is a guard over nodes of the type Clock contained in the left-hand side graph L,*
- $r_C \subseteq CN(L)$ *is a set of nodes of the type Clock contained in the left-hand side graph L to be reset.*

In the following we give a short example for a probabilistic timed rule.

Example 7 [Probabilistic Timed Graph Transformation Rule]. As an example we model the failure of a hardware node using a probabilistic timed rule *fail*, which has two right-hand sides for the case where the node fails with the probability of 10% or not with the probability of 90%. The adjacent clock is used to ensure that the probabilistic timed rule is executed not more than every two time units by resetting the clock during each application and by using a guard that requires the clock to have a value greater than or equal two to capture that the hardware node can fail for not modeled external requests with a minimal arrival time greater than or equal two time units. The underlying type graph for this rule is depicted in the picture (a) below. Formally, the rule is given by $fail = (L_1, P_1, \mu_1, \phi_1, r_{C_1})$ with L_1 as given in the picture (b) to the left, $P_1 = \{\rho_1^0, \rho_1^1\}$ with $rhs(\rho_1^0)$[4] and $rhs(\rho_1^1)$ as given in the picture (b) to the right, $\mu_1 = \{(\rho_1^0, 0.1), (\rho_1^1, 0.9)\}$, $\phi_1 = (c \geq 2)$, and $r_{C_1} = \{c\}$.

---

[4] $rhs(\rho)$ denotes the right-hand side of the rule ρ.

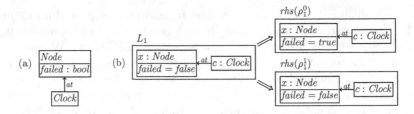

Invariants, as a central concept of PTA, are given for PTGTSs in the form of conditions over clocks that are checked to be satisfied for a given configuration.

Definition 8 (Probabilistic Timed Graph Transformation Invariant).
$\Theta = (L, \phi)$ *is a probabilistic timed graph transformation invariant if*

- *L is a graph,*
- *$\phi \in \Phi(CN(L))$ is an invariant formula over nodes of the type Clock contained in the graph L.*

For atomic propositions, which are used in the context of the PTCTL logic, we make use of the same kind of conditions as for invariants. For this reason we formally denote an atomic proposition also by $\Theta = (L, \phi)$. The atomic propositions are checked to set the appropriate labels to the PTGTS configurations.

In the following we define PTGTSs comprising the notions introduced above. For a concrete example of a PTGTS see Sect. 4.

Definition 9 (Probabilistic Timed Graph Transformation System).
$S = (TG, G_0, v_0, \Pi, I, AP, prio)$ *is a probabilistic timed graph transformation system (PTGTS) if*

- *TG is a finite type graph including the type node Clock,*
- *G_0 is a finite initial graph over TG,*
- *$v_0 : CN(G_0) \to \mathbb{R}$ is the initial clock valuation assigning the clock value 0 to every node of the type Clock in G_0,*
- *Π is a finite set of probabilistic timed rules,*
- *I is a finite set of probabilistic timed invariants,*
- *AP is a finite set of probabilistic timed atomic propositions,*
- *$prio : \Pi \to \mathbb{N}$ is a priority function assigning a priority to each rule[5].*

As a next step we define when a PTGTS configuration consisting of a graph and a current clock valuation satisfies some invariant.

Definition 10 (Probabilistic Timed Invariant Satisfaction). *Let $S = (TG, G_0, v_0, \Pi, I, AP, prio)$ be a PTGTS, $\Theta = (L, \phi) \in I$, G be a graph typed over TG, and $v \in \mathcal{V}(CN(G))$ be a clock valuation for nodes of the type Clock in G. Then $(G, v) \models \Theta$ if for every injective match $m : L \to G$ with $m = (m_V, m_E)$ it holds that $v \circ m_V \models \phi$.*

---

[5] For the priority function it holds that the higher the number assigned to a rule the higher is the priority of the rule.

Not every configuration of a PTGTS reached by rule based structure adaptation is valid since the invariants of the PTGTS need to be considered, too.

Definition 11 (Probabilistic Timed Graph Transformation State). *Let* $S = (TG, G_0, v_0, \Pi, I, AP, prio)$ *be a PTGTS. Then* $q = (G, v)$ *is a probabilistic timed graph transformation state (also called configuration) of* S *(written* $q \in states(S)$*) if*

- *G is a graph typed over TG,*
- *$v \in \mathcal{V}(CN(G))$ is a clock valuation for nodes of the type Clock in G,*
- *$(G, v) \models \Theta$ for each $\Theta \in I$.*

The behavior of PTGTSs is defined by the probabilistic timed graph transformation steps. We distinguish here similarly to PTA two kinds of steps: timed steps increasing the clock values by the time elapsed and transition steps allowing to switch a configuration under certain conditions as defined below. For the transition steps we ensure that the guard of the rule is satisfied by the current clock valuation and match, that no transition step with higher priority can be executed, and that all steps in the selected probability distribution μ are enabled. Furthermore, considering a rule ρ with non-zero probability $\mu(\rho)$, we define the single rule single step relation based on the expected DPO transformation step and ensure that the clock valuations of the source and target states are compatible also enforcing the clock resets of the rule.

Definition 12 (Probabilistic Timed Graph Transformation Step). *Let* $S = (TG, G_0, v_0, \Pi, I, AP, prio)$ *be a PTGTS. Then the* single step relation *is given as follows:*

- ***Timed Step:*** $(G, v) \xrightarrow{\delta} {}^{\mathrm{PTGTS}}_S (G, v + \delta)$ *if* $\delta > 0$ *and for each* δ' *it holds that* $0 \leq \delta' \leq \delta$ *implies that* $(G, v + \delta') \in states(S)$.
- ***Transition Step:*** $(G_1, v) \xrightarrow{R, \rho, m} {}^{\mathrm{PTGTS}}_S (G_2, v')$ *if*
  - $R = (L, P, \mu, \phi, r_C) \in \Pi$ *is a probabilistic timed rule,*
  - $m : L \to G_1$ *with* $m = (m_V, m_E)$ *is an injective match,*
  - $v \circ m_V \models \phi$,
  - $\rho \in P$ *is a transformation rule with non-zero probability* $\mu(\rho) > 0$,
  - $\nexists G_2', v'', R', \rho', m'$ *such that* $(G_1, v) \xrightarrow{R', \rho', m'} {}^{\mathrm{PTGTS}}_S (G_2', v'')$ *and* $prio(R') > prio(R)$,
  - $(G_1, v) \xrightarrow{R, \rho, m} {}^{\mathrm{PTGTS}}_S (G_2, v')$,
  - $\forall \rho' \in P \setminus \{\rho\}$ *such that* $\mu(\rho') > 0$ *there is a graph* G_2' *such that* $(G_1, v) \xrightarrow{R, \rho', m} {}^{\mathrm{PTGTS}}_S (G_2', v')$,

  where $(G_1, v) \xrightarrow{R, \rho, m} {}^{\mathrm{PTGTS}}_S (G_2, v')$ *is the* single rule single step relation *if*
  - $(G_1, v), (G_2, v') \in states(S)$,
  - $\rho = (L \xleftarrow{l} K \xrightarrow{r} R)$ *is a graph transformation rule,*
  - $(1) + (2)$ *is a DPO diagram for the transformation step* $G_1 \Longrightarrow \rho, m G_2$,

- *clock valuation functions* $v : CN(G_1) \to \mathbb{R}$ *and* $v' : CN(G_2) \to \mathbb{R}$ *are compatible, i.e.,* $\forall X \in CN(G_1).$ $(\forall Y \in CN(D). (l'_V(Y) = X) \Rightarrow (v'(r'_V(Y)) = v[m_V(r_C) := 0](X)))^6$,
- *the clock value 0 is assigned to all created nodes of the type Clock, i.e.,* $\forall Z \in CN(G_2) \backslash r'_V(CN(D)). v'(Z) = 0.$

$$
\begin{array}{ccccc}
L & \xleftarrow{\ l\ } & K & \xrightarrow{\ r\ } & R \\
\downarrow{\scriptstyle m} & (1) & \downarrow & (2) & \downarrow{\scriptstyle k} \\
G_1 & \xleftarrow{\ l'\ } & D & \xrightarrow{\ r'\ } & G_2
\end{array}
\qquad
\begin{array}{ccccc}
CN(L) & \xleftarrow{\ l_V\ } & CN(K) & \xrightarrow{\ r_V\ } & CN(R) \\
\downarrow{\scriptstyle m_V} & = & \downarrow & = & \downarrow{\scriptstyle k_V} \\
CN(G_1) & \xleftarrow{\ l'_V\ } & CN(D) & \xrightarrow{\ r'_V\ } & CN(G_2)
\end{array}
$$
$$
\begin{array}{ccc}
& \searrow^{v} \quad \swarrow^{v'} & \\
& \mathbb{R} &
\end{array}
$$

In our subsequent translation of PTGTSs into the corresponding PTSs we identify configurations of PTGTSs up to isomorphism. For this purpose we now introduce such isomorphisms.

Definition 13 (Isomorphisms on States of PTGTSs). *Let S be a PTGTS and $(G_1, v_1), (G_2, v_2) \in states(S)$.*
    Then $(G_1, v_1) \cong (G_2, v_2)$ if

$$
\begin{array}{ccc}
CN(G_1) & \xrightarrow{\ i_V\ } & CN(G_2) \\
& \searrow_{v_1} \ = \ \swarrow_{v_2} & \\
& \mathbb{R} &
\end{array}
$$

- *$i : G_1 \to G_2$ with $i = (i_V, i_E)$ is an isomorphism,*
- *$\forall X \in CN(G_1)$ it holds that $v_1(X) = v_2(i_V(X))$.*

Analogously to PTA we provide a PTS for every PTGTS and, hence, allow for a comparison of PTA and PTGTSs by comparing their semantics in the sense of the corresponding PTSs.

Definition 14 (Induced PTS for a PTGTS). *Let $S = (TG, G_0, v_0, \Pi, I, AP, prio)$ be a PTGTS. Then $\mathcal{M}_S = (Q_S, Steps_S, L_S)$ is the induced PTS if*

- $Q_S = \{[(G, v)]_\cong \mid (G_0, v_0) \xrightarrow{*}{}_S^{\mathrm{PTGTS}} (G, v)\}$,
- $L_S([(G, v)]_\cong) = \{\Theta \in AP \mid (G, v) \models \Theta\}$,
- $Steps_S([(G, v)]_\cong) = \{(\delta, \nu) \mid (G, v) \xrightarrow{\delta}{}_S^{\mathrm{PTGTS}} (G', v') \wedge \nu([(G', v')]_\cong) = 1\} \cup \{(0, \nu) \mid R = (L, P, \mu, \phi, r_C) \in \Pi \wedge \rho \in P \wedge (G, v) \xrightarrow{R, \rho, m}{}_S^{\mathrm{PTGTS}} (G', v')\}$ *where* $\nu([(\overline{G}, \overline{v})]_\cong) = \sum_{\rho' \in P : (G,v) \xrightarrow{R, \rho', m}{}_S^{\mathrm{PTGTS}} (G'', v'') \cong (\overline{G}, \overline{v})} \mu(\rho').$

In the induced PTS we consider configurations up to isomorphism using their equivalence classes and derive the labeling of configurations by evaluating the atomic propositions of the PTGTS. For the step relation we need to collate PTGTS steps with common target when constructing the probability distribution ν to ensure well-definedness.

We furthermore employ negative application conditions (*NAC*s) [12] and attributes for PTGTSs. They allow to increase the descriptive expressiveness of the rules and can be added straightforwardly to the presented formalization.

---

^{6} For morphisms between clocks we omit the restricted notation $f_V|_{CN(G_1)} : CN(G_1) \to CN(G_2)$ and use the unrestricted notation $f_V : CN(G_1) \to CN(G_2)$ to simplify the representation.

4   Modeling

To support modeling and, subsequently, analysis of PTGTS models with their probabilistic and timed behavior, we extended the existing support of HEN-SHIN [8] for PGTSs [16]. Analogously to PGTSs, the elements and links between the elements are captured by an EMF model represented as a class diagram (the type graph of the PTGTS as given in Fig. 1a). In addition, we require a *Clock* element to be present in the EMF model to enable the modeling of timed behavior. The probabilistic choices are modeled as for PGTSs with multiple HENSHIN transformation rules with the same name and the same left-hand side (e.g., depicted in Figs. 2d and e for the rule *connect*). To support the modeling of real-time behavior, we associate clock guards (CG), clock invariants (CI), and clock resets (CR) to the rules via corresponding annotations added to the property list of the GTS in HENSHIN. Consequently, the HENSHIN model includes all details of a PTGTS model. Since HENSHIN does not include rules' annotations in their visual representation we label the rules in this paper (e.g., in Fig. 2) with (CG); (CI); (CR) where void elements are represented by —.

Syntactically, a rule $L \xleftarrow{l} K \xrightarrow{r} R$ in HENSHIN is given by a single graph annotated with specific stereotypes. The stereotypes *«preserve»*, *«delete»*, and *«create»* correspond to the elements of K, L without K, and R without K, respectively. The stereotype *«forbid»* is used to specify *NAC*s and can be parametrized as in *«forbid#n»* for $n \in \mathbb{N}$ to distinguish between multiple *NAC*s.

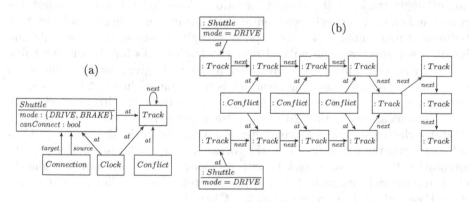

Fig. 1. Type graph of the shuttle scenario (a) and topology with 3 conflict nodes (b)

As a running example, we model a scenario inspired by the RailCab project [24], where a service choreography coordinates the movement of shuttles on tracks, as a PTGTS using HENSHIN. The type graph and the initial track topology are given in Fig. 1. In the context of our scenario, tracks are connected to the adjacent tracks by *next* edges. Shuttles are located on tracks, which is represented by *at* edges. Shuttles can move forward on tracks being in the *DRIVE* mode or can initiate emergency brakes changing to the *BRAKE* mode to avoid collisions. To avoid collisions shuttles can also communicate and establish connections. A connection is associated with the leading shuttle via a *target* edge and with the following shuttle via a *source* edge. The connection attempt between two shuttles may fail, but it can be repeated after both involved shuttles moved one track forward. This aspect is expressed by the shuttle's attribute *canConnect*. Two connected shuttles are allowed to be at the same track without being involved in a collision. In the initial topology parallel tracks leading to the same track after one, two or more successor tracks are marked by *conflict* nodes. Two shuttles can try to establish a connection if they are on tracks connected by a *conflict* node. Another reason for communication of the shuttles is the reduction of energy consumption. For this reason shuttles can form convoys also establishing a connection. We also equip shuttles and tracks with *clocks* needed for time measurement only to be able to control the time for rule applications. Note that we do not depict the nodes of the type *Clock* in the rules explicitly to keep the rule representation concise but use the annotation $e.c$ to refer to the clock c linked to some element e.

The behavior of the shuttle scenario is modeled in HENSHIN using the following probabilistic timed rules. Shuttles can drive alone or can build convoys to reduce the energy consumption. The rule *driveAlone* (see Fig. 2a) allows a shuttle that is leading a convoy or a shuttle driving without a convoy to move forward if there are no shuttles located at the track after the subsequent track and any of that track's predecessor tracks. The four *driveConvoy* rules (see [21, Chap. 4]) allow a shuttle to follow a leading shuttle depending on the layout of single tracks in the current situation. Following the aim to reflect the real-time behavior properly, we require that moving on a single track can take between 3 and 4 min, which we express using the corresponding guards and invariants, respectively, formulated over the track clocks for the driving rules. For the rule *driveAlone*, the corresponding guard is given by the annotation $t1.c \geq 3$ and the corresponding invariant is not depicted in Fig. 2a but is given by $\Theta_{driveAlone} = (lhs(driveAlone), t1.c \leq 4)$ for a track $t1$ with its clock c. To be able to measure the time spent on a track properly, we reset the clock of a track to which a shuttle is moving after applying one of the driving rules. Considering again the rule *driveAlone*, the corresponding clock reset is given by the annotation $t2.c' = 0$ for a track $t2$ with its clock c'. The corresponding guards, clock resets, and invariants for *driveConvoy* rules look similar to that of the *driveAlone* rule and can be found in [21, Chap. 4].

Shuttles may connect with each other to create convoys and to prevent collisions. Figures 2d and e depict the probabilistic timed rule *connect* allowing two driving shuttles located at parallel critical tracks to communicate and to create a convoy. The *NAC*s of the rule express intuitively that there must not already exist a connection between the two considered shuttles, the shuttle chosen as the leader must not be leading another convoy, and the shuttle chosen as the following shuttle must not be following another shuttle. Since a connection request can be lost after its sending, the rule *connect* has two different right-hand sides representing on the one hand, the case that the connection is established successfully

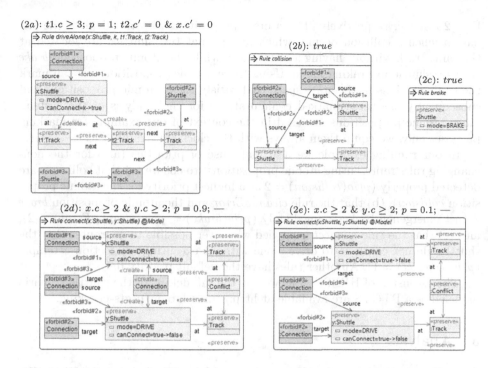

Fig. 2. Rules and atomic propositions of the shuttle PTGTS

as depicted in Fig. 2d (which happens with the probability of 90%) and on the other hand the situation that the connection request has been lost as depicted in Fig. 2e (which occurs with the probability of 10%). We assume furthermore that each communication attempt takes at least two minutes for each shuttle and model this behavior by using the corresponding guard $x.c \geq 2$ & $y.c \geq 2$ (for two shuttles x, y and their respective clocks c) for communicating shuttles in both basic rules *connect* (see Figs. 2d and e). Moreover, since a communication attempt can be repeated only after each of the shuttles has moved one track forward, we reset a clock of a shuttle each time it has used the rule *driveAlone* after a communication attempt, which is formulated by the corresponding clock reset $x.c' = 0$ of the rule *driveAlone* for a shuttle x with its clock c'.

In the case if no connection attempt was successful, a shuttle, which is driving behind another one, has to brake to avoid a collision, if both shuttles come too close to each other. The two situations in which a shuttle has to initiate an emergency brake are depicted in [21, Chap. 4].

After a shuttle has changed into braking mode, the system detects the safety-critical situation and terminates using the rule *cleanupError*. Otherwise, if one of both shuttles can reach the end state of the track system successfully, the system terminates using the rule *cleanupOk*. Both rules are given in [21, Chap. 4].

In the context of our shuttle scenario, we consider two atomic propositions *collision* and *brake* modeled as non-changing rules in HENSHIN as depicted in

Figs. 2b and c, respectively. The atomic proposition *collision* depicts the situation when a collision occurs, which means that two shuttles are located at the same track without having a connection, while the atomic proposition *brake* shows a shuttle in braking mode. Using these atomic propositions we can mark the configurations in the state space that satisfy these atomic propositions and subsequently use these marked configurations for the analysis of the system. Since atomic propositions are used in the context of our example for marking purposes only, we equip them always with the clock constraint *true*.

In our running example we also make use of priorities (also for the non-changing rules representing atomic propositions) to ensure (a) that collisions are detected properly ($prio(collision) = 2$ as a highest priority for the atomic proposition *collision*), (b) that the rule *cleanupError* and the atomic proposition *brake* detect emergency brakes immediately ($prio(brake) = prio(cleanupError) = 2$), and (c) that the rule *connect* is applied whenever possible to guarantee that the shuttles do not continue to drive alone without executing connection attempts ($prio(connect) = 1$). All other rules have the default priority 0.

The extension of HENSHIN as well as the full details of the shuttle scenario modeled as a PTGTS are available at http://mde-lab.de/ptgts-checking.

5   Analysis

As outlined in the previous section, we can model PTGTSs using the HENSHIN tool. To analyze a PTGTS model, we can chain together the capabilities of the HENSHIN tool for GTSs and the PRISM model checker for PTA.

In *Step 1* we use the capability of HENSHIN to generate the state space of a GTS by considering the probabilistic choices between the different right-hand sides of probabilistic timed rules as if they were nondeterministic and ignoring clock guards, clock resets as well as clock invariants. In *Step 2* we extend our mapping from PGTSs to PA [16] to be able to convert the state space generated for the PTGTS into the corresponding PTA. In this step we replace the nondeterministic choice between the different right-hand sides of probabilistic timed rules by probabilistic transitions, including clock guards and clock resets as well as adding atomic propositions and the set of clock invariants, which must hold for all valid states of the system. The PTA generated in this way has an input format of PRISM allowing to verify PTCTL properties by computing the corresponding minimum and maximum probabilities. Finally, in *Step 3* we can model check the resulting PTA with PRISM according to the properties of interest for the PTGTS. Note that this tool chain can only be used in practice, if the state space generation using HENSHIN terminates in *Step 1* and results in a finite state space of moderate size, which PRISM is capable to analyze.

For the shuttle PTGTS described in the previous section, we executed several experiments by using this outlined tool chain.

In our first experiment we determined using PRISM the states generated by HENSHIN that remain reachable with non-zero probability when considering the timing behavior. Thereby we exclude many of the calculated traces from the

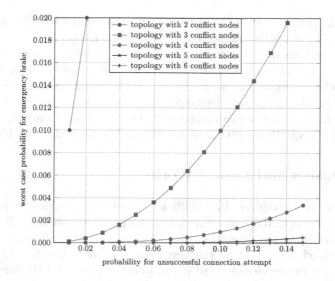

Fig. 3. Visualization for the probability of an emergency brake

further analysis since they do not satisfy the corresponding time constraints. For the shuttle example PRISM detects for the topologies with 2–6 conflict nodes 53.1%, 58.0%, 62.0%, 64.3%, and 65.8% of the states generated by HENSHIN to be non-reachable. The results of this experiment are plotted in [21, Chap. 5].

In our second experiment we analyzed whether the considered shuttle scenario can under any circumstances exhibit a collision. In fact, collisions between shuttles cannot occur due to the nature of the contained transformation rules, which ensure that emergency brakes are applied if necessary. This can be verified already in HENSHIN after *Step 1* by using the atomic proposition *collision* (see Fig. 2b) with the highest priority detecting a collision by marking configurations in the state space where the atomic proposition can be matched. In HENSHIN we then observe that this atomic proposition labels no state and, hence, no additional analysis using PRISM is required.

Finally, in our third experiment we verified using PRISM the maximal probability with which the described shuttle system executes an emergency brake. For this reason we generated first using HENSHIN in 11.8–17.5 s state spaces with 81–269 states for the different topologies with 2–6 conflict nodes, respectively. The property of interest is represented in PRISM notation by $Pmax =? [F$ *"brake"*$]$ for the atomic proposition *brake* given in Fig. 2c and the exists-eventually-operator F. In Fig. 3 we show for topologies with 2–6 conflict nodes, which also determine the initial distance of the shuttles to the critical track element, how the corresponding maximal probabilities depend on the likelihood of the successful connection establishment. As expected, the lower the probability for a non-successful connection attempt (x-axis) the lower the maximal probability for emergency brake execution (y-axis), which is the worst case scenario[7]. The computation of the probability values using PRISM required for topologies with 2–6 conflict nodes 0.3–181.7 s, respectively.

We can conclude that our running example modeled as a PTGTS behaves as desired because (a) collisions are avoided altogether and (b) the worst case probabilities for emergency brake can be controlled using the number of conflict nodes based on the likelihood of unsuccessful connection attempts. Our experiments demonstrate that we can analyze PTGTSs by employing the explained tool chain of HENSHIN operating on GTSs and PRISM analyzing PTA.

6   Conclusion and Future Work

In this paper we introduced Probabilistic Timed Graph Transformation Systems (PTGTSs) as a high-level description language supporting all the necessary aspects of structure dynamics, timed behavior, and probabilistic behavior that we identified as relevant for the next generation of embedded real-time systems employing the service-oriented paradigm. We presented the formal model of PTGTSs and outlined a mapping of PTGTS models employing the HENSHIN tool to probabilistic timed automata (PTA) such that the PRISM model checker can be used to analyze PTGTS models with respect to PTCTL properties.

As a future work we plan to provide a specification formalism operating on PTGTS to be able to state more complex properties on the structure dynamics, timed behavior, and probabilistic behavior of the given PTGTS model in a coherent way. Such an extension is then to be included in the mapping of PTGTS to PTA to allow for their automated verification using PRISM.

References

1. Agha, G., Meseguer, J., Sen, K.: PMaude: rewrite-based specification language for probabilistic object systems. ENTCS **153**, 213–239 (2006)
2. Alur, R., Courcoubetis, C., Dill, D.L.: Model-checking in dense real-time. Inf. Comput. **104**(1), 2–34 (1993)
3. Alur, R., Dill, D.L.: A theory of timed automata. TCS **126**(2), 183–235 (1994)
4. Becker, B., Giese, H.: On safe service-oriented real-time coordination for autonomous vehicles. In: ISORC 2008, pp. 203–210 (2008)
5. Bellman, R.: A Markovian decision process. Indiana Univ. Math. J. **6**, 679–684 (1957)
6. Bouyssounouse, B., Sifakis, J. (eds.): Embedded Systems Design: The ARTIST Roadmap for Research and Development. Springer, Heidelberg (2005)
7. Daws, C., Olivero, A., Tripakis, S., Yovine, S.: The tool KRONOS. In: Alur, R., Henzinger, T.A., Sontag, E.D. (eds.) HS 1995. LNCS, vol. 1066, pp. 208–219. Springer, Heidelberg (1996). doi:10.1007/BFb0020947
8. The Eclipse Foundation: EMF Henshin (2013). http://www.eclipse.org/modeling/emft/henshin

---

[7] The range of the considered probabilities for non-successful connection attempts has been taken from [20] where for close range communication and high data rates an error rate of at most 13% has been observed for wireless communication.

9. Ehrig, H., Ehrig, K., Prange, U., Taentzer, G.: Fundamentals of Algebraic Graph Transformation. Springer, Heidelberg (2006)
10. Giese, H.: Modeling and verification of cooperative self-adaptive mechatronic systems. In: Kordon, F., Sztipanovits, J. (eds.) Monterey Workshop 2005. LNCS, vol. 4322, pp. 258–280. Springer, Heidelberg (2007). doi:10.1007/978-3-540-71156-8_14
11. Gyapay, S., Varró, D., Heckel, R.: Graph transformation with time. Fundam. Inf. **58**, 1–22 (2003)
12. Habel, A., Heckel, R., Taentzer, G.: Graph grammars with negative application conditions. Fundam. Inf. **26**(3,4), 287–313 (1996)
13. Heckel, R., Lajios, G., Menge, S.: Stochastic graph transformation systems. Fundam. Inf. **74**, 63–84 (2006)
14. Henzinger, T.A., Kupferman, O.: From quantity to quality. In: Maler, O. (ed.) HART 1997. LNCS, vol. 1201, pp. 48–62. Springer, Heidelberg (1997). doi:10.1007/BFb0014712
15. Kastenberg, H., Rensink, A.: Model checking dynamic states in GROOVE. In: Valmari, A. (ed.) SPIN 2006. LNCS, vol. 3925, pp. 299–305. Springer, Heidelberg (2006). doi:10.1007/11691617_19
16. Krause, C., Giese, H.: Probabilistic graph transformation systems. In: Ehrig, H., Engels, G., Kreowski, H.-J., Rozenberg, G. (eds.) ICGT 2012. LNCS, vol. 7562, pp. 311–325. Springer, Heidelberg (2012). doi:10.1007/978-3-642-33654-6_21
17. Kwiatkowska, M., Norman, G., Parker, D.: PRISM 4.0: verification of probabilistic real-time systems. In: Gopalakrishnan, G., Qadeer, S. (eds.) CAV 2011. LNCS, vol. 6806, pp. 585–591. Springer, Heidelberg (2011). doi:10.1007/978-3-642-22110-1_47
18. Kwiatkowska, M., Norman, G., Sproston, J., Wang, F.: Symbolic model checking for probabilistic timed automata. Inf. Comput. **205**, 1027–1077 (2007)
19. Kwiatkowska, M.Z., Norman, G., Segala, R., Sproston, J.: Automatic verification of real-time systems with discrete probability distributions. TCS **282**(1), 101–150 (2002)
20. Lan, K., Chou, C., Jin, D.: The effect of 802.11 a on DSRC for ETC communication. In: WCNC 2012, pp. 2483–2487 (2012)
21. Maximova, M., Giese, H., Krause, C.: Probabilistic timed graph transformation systems. Technical report 118, Hasso-Plattner Institute at the University of Potsdam (2017)
22. Neumann, S.: Modellierung und Verifikation zeitbehafteter Graphtransformationssysteme mittels Groove. Master's thesis, University of Paderborn (2007)
23. Ölveczky, P.C., Meseguer, J.: Semantics and pragmatics of real-time Maude. HOSC **20**, 161–196 (2007)
24. RailCab homepage. http://www.railcab.de
25. Schäfer, W., Wehrheim, H.: The challenges of building advanced mechatronic systems. In: FOSE 2007, pp. 72–84 (2007)
26. Segala, R.: Modeling and verification of randomized distributed real-time systems. Ph.D. thesis, Massachusetts Institute of Technology (1996)
27. Sen, K., Viswanathan, M., Agha, G.A.: VESTA: a statistical model-checker and analyzer for probabilistic systems. In: QEST 2005, pp. 251–252 (2005)

Model Transformation and Tools

Leveraging Incremental Pattern Matching Techniques for Model Synchronisation

Erhan Leblebici[1]([✉]), Anthony Anjorin[2], Lars Fritsche[1], Gergely Varró[1], and Andy Schürr[1]

[1] Technische Universität Darmstadt, Darmstadt, Germany
{erhan.leblebici,lars.fritsche,
gergely.varro,andy.schurr}@es.tu-darmstadt.de
[2] Universität Paderborn, Paderborn, Germany
anthony.anjorin@uni-paderborn.de

Abstract. Triple Graph Grammars (TGGs) are a declarative, rule-based approach to model synchronisation with numerous implementations. TGG-based approaches derive typically a set of *operational* graph transformations from direction-agnostic TGG rules to realise *model synchronisation*. In addition to these derived graph transformations, however, further runtime analyses are required to calculate the consequences of model changes in a synchronisation run. This part of TGG-based synchronisation is currently manually implemented, which not only increases implementation and tool maintenance effort, but also requires tool or at least approach-specific proofs for correctness. In this paper, therefore, we discuss how *incremental* graph pattern matchers can be leveraged to simplify the runtime steps of TGG-based synchronisation. We propose to outsource the task of calculating the consequences of model changes to an underlying incremental pattern matcher. As a result, a TGG-based synchroniser is reduced to a component reacting solely to appearing and disappearing matches. This abstracts high-level synchronisation goals from low-level details of handling model changes, providing a viable and unifying foundation for a new generation of TGG tools.

1 Introduction and Motivation

Bidirectional model synchronisation is a current challenge that is becoming increasingly relevant in numerous domains [4]. In our context, bidirectional model synchronisation refers to the task of keeping two models (called *source* and *target*) consistent by propagating changes (called *deltas*) applied to one of the models, i.e., by executing a forward or backward transformation to restore consistency. The task of implementing an *incremental* synchroniser with clear and precise semantics is non-trivial. In this paper, an incremental forward[1] synchroniser takes the old target model into account when propagating source deltas

---

[1] In the entire paper, symmetric statements that hold analogously in both forward and backward directions are only formulated in the forward direction for brevity.

© Springer International Publishing AG 2017
J. de Lara and D. Plump (Eds.): ICGT 2017, LNCS 10373, pp. 179–195, 2017.
DOI: 10.1007/978-3-319-61470-0_11

(and does not create the target model from scratch). Bidirectional transformation (bx) languages address this task via diverse techniques [4].

Triple Graph Grammars (TGGs) [16] as a bx language, represent a declarative, rule-based approach to model synchronisation based on the mature field of graph transformations [5]. TGG rules are direction-agnostic, describing how consistent pairs of source and target models can be created simultaneously. TGG-based model synchronisation typically involves compile time and runtime subtasks: At compile time, operational (forward and backward) graph transformation rules are derived from TGG rules. At runtime, consequences of deltas with respect to applying these operational rules are calculated, and consistency is restored by revoking invalidated rule applications from former runs and performing new ones. This step ideally guarantees *correctness*, i.e., that the resulting pair of a source and target model can be created by applying the TGG rules.

All state-of-the-art TGG-based synchronisation frameworks we are aware of [7,8,10,11,13,15] address the runtime step (i) in a simple but non-scalable manner, starting each time from scratch and considering the entire models [10], or (ii) by providing auxiliary dependency analyses over the source (target) model [13,15] or correspondences [7], or (iii) by applying practically useful but as yet informal heuristics without proofs of correctness for all possible cases [8,11]. Our observation is that the complexity in addressing the runtime steps of a TGG-based synchronisation is accidental and is caused by entangling high-level incremental propagation strategies with low-level details of how deltas and their transitive consequences must be handled efficiently and correctly.

Incremental pattern matching techniques (e.g., [6]) provide a viable means of monitoring all matches of a given set of patterns in a host graph and thus can observe and report consequences of deltas. Although this naturally addresses the runtime requirements of TGG-based model synchronisation, incorporating incremental pattern matchers into TGGs has not yet been analysed up until now. Our contribution is, therefore, to integrate incremental pattern matching and TGG-based model synchronisation. Our aim is to provide a formal foundation for a new generation of TGG tools that can now leverage available incremental graph pattern matching tools [17,18]. We are able to reduce a TGG-based synchroniser to a relatively simple component that reacts to invalidated or available rule applications reported by its underlying incremental pattern matcher.

The paper is structured as follows: We present in Sect. 2 a compact but non-trivial synchronisation scenario and discuss the diverse delta propagation strategies. A novel concept for TGG-based synchronisers making use of incremental pattern matching techniques is presented intuitively in Sect. 3, and formalised in Sect. 4 with correctness arguments, shaping our main contribution. Related approaches and future work are discussed in Sects. 5 and 6, respectively.

2 Running Example and Preliminaries

A TGG specification consists of a *schema* and a set of *rules*. A schema is a triple of metamodels representing the abstract syntax of source, target, and correspondence models used for mappings between source and target elements.

Example: Our running example is *Ecore2HTML*.[2] The scenario addressed is the general usage of models for *editable* report generation; we only focus here on packages and their corresponding folders and files. The TGG schema for our running example is depicted in Fig. 1a. Source models are hierarchies of packages and target models (representing reports) are hierarchies of folders where folders may contain files. The single correspondence type (P2F) connects packages to folders. With respect to this TGG schema, a correctly typed model triple is depicted in Fig. 1b consisting of three packages p1, p2, and p3 as well as three corresponding folders f1, f2, and f3. The outermost folders f1 and f3 additionally contain a file fe1 and fe3, respectively.

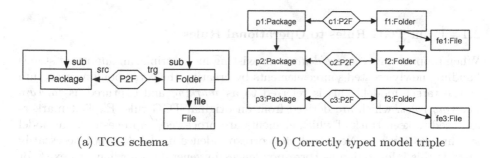

(a) TGG schema   (b) Correctly typed model triple

Fig. 1. Schema and typed model triple

We use a compact syntax to represent TGG rules, merging both the precondition L and the postcondition R together in a single diagram. The elements in L (also referred to as context elements of the rule) are black, while elements in $R\backslash L$ (also referred to as created elements of the rule) are green with a ++ markup. Edges that are incident to a created node do not have explicit ++ markup as they must also obviously be created. A model triple is *consistent* with respect to a TGG, if it can be created by applying the rules of the TGG.

Example: Intuitively, what we want to specify is that packages correspond to folders and, additionally, that outermost packages require an extra file containing project-level documentation in their corresponding folder. To achieve this with TGGs, we need two rules: `PackageDocRule` (we shall also refer to this as R1) depicted in Fig. 2a, creates a package p, together with a corresponding folder f with a file `fe`. The created package and folder are also connected with a correspondence link c. R1 has no context elements and can thus be applied to the empty triple. `SubPackageDocRule` (R2) depicted in Fig. 2b, requires a package p and a corresponding folder f (note how the correspondence link c is used to enforce this), and extends the package and folder hierarchies by creating a new sub-package p' and subfolder f', connected via the correspondence link c'. In contrast to R1, rule R2 has context elements and can only be applied to

---

[2] The entire synchronisation scenario including our excerpt is documented in the bx example repository at http://bx-community.wikidot.com/examples:ecore2html.

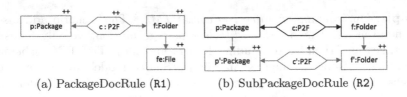

(a) PackageDocRule (R1)   (b) SubPackageDocRule (R2)

Fig. 2. TGG rules for the running example

extend an existing model triple. In this sense, the triple depicted in Fig. 1b can be created with these rules and is thus consistent with respect to the TGG.

2.1 From TGG Rules to Operational Rules

When propagating a *source delta* in an existing model triple, an important step is handling newly created source elements by creating the corresponding structure in the target model. This is referred to as *marking* and captured as *forward marking rules*, which are derived from an original TGG rule. Explicit markers are used to keep track of which elements are processed/unprocessed in a model synchronisation run. Note that the correspondence model does not necessarily provide this information as correspondences in general do not have to exist for each element or multiple correspondences might exist for the same element.

Intuitively, the forward marking rule of a TGG rule does not create any source element but requires them, marks all created elements of its respective TGG rule, and requires that all context elements of the TGG rule be marked (e.g., by former applications of forward marking rules). Additionally, *Negative Application Conditions* (NACs) are used to ensure that source elements (the input elements in case of a forward synchronisation) are marked only once as they would be created once by the original TGG rules. Such NACs are referred to as *marker NACs*. Finally, an optional set of *filter NACs* is used to avoid invalid rule applications that would lead to a state where, e.g., certain edges can no longer be marked (such NACs are constructed automatically via static analysis techniques [9]).

Example: The forward marking rules for our running example are depicted in Fig. 3. All elements belonging to a NAC are depicted blue and crossed-out. For presentation purposes, markers are denoted by circles that are connected to elements (nodes and edges). The forward marking rule derived from R1 (Fig. 3a) creates an R1 marker connected to the package p and all created elements c, f and fe. The package p is demanded as context that must not already be marked (via the blue, crossed-out marker connected to it). If this forward marking rule were ever used to mark a sub-package p, it would be impossible to ever mark the incoming sub edge to p, as there does not exist a rule that creates an incoming sub edge to an existing package. Hence, a filter NAC is used to forbid the presence of such edges, i.e., the forward marking rule derived from R1 can only mark outermost packages (as the original TGG rule R1 can only create outermost packages). In the forward marking rule derived from R2 (Fig. 3b), the context

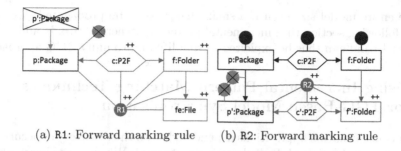

(a) R1: Forward marking rule   (b) R2: Forward marking rule

Fig. 3. Derived marking rules for the running example

elements of the original TGG rule are additionally required as already marked (no matter whether by the same or different markers).

2.2 Delta Propagation via Operational Rules

Given a consistent model triple and a source delta, the main task of forward synchronisation is to detect invalidated and available applications of forward marking rules. Invalidated applications (e.g., due to deleted context elements) must be *revoked* by deleting their created correspondence and target elements as well as obsolete markers. Conversely, available applications of forward marking rules lead to new correspondence and target elements with new markers.

Example: A simple source delta in our example is given by creating (deleting) a package, leading to an available (invalidated) application of the forward marking rule of R1 or R2 (depending on whether the package is an outermost package or not). A non-trivial source delta is given by creating a sub edge such that a former outermost package becomes a child package, e.g., creating a sub edge from p2 to p3 in Fig. 1b. In this case, an application of the forward marking rule of R1 becomes invalid (the filter NAC is violated as the outermost package p3 now becomes a child package). After deleting the obsolete marker of p3 (and the corresponding target elements), an application of the forward marking rule of R2 becomes available, i.e., p3 can now be re-marked as a subpackage.

Existing TGG approaches differ from each other mainly concerning how invalid or available applications of operational rules are detected. In *precedence-driven* approaches [13,15], an auxiliary precedence analysis between model elements is performed (and maintained) to determine which model elements are potentially affected by deletions or creations of others. Alternatively, this analysis is performed between correspondences [7] (affected correspondences are calculated for a given source delta). Such hand-crafted analyses, however, either overestimate the actual dependencies as dependencies are retrieved at the type level [13], or underestimate them relying on additional information via user-interaction [15] or special correspondences [7]. A completely different strategy is to re-mark an existing triple from scratch and to complement missing markers in a final step [10]. This, however, makes the synchronisation process dependent on the entire model size even if a small change is to be propagated. We argue in the following section that incremental pattern matchers naturally address the same tasks and can thus be exploited to simplify and to unify TGG approaches.

3 Using Incremental Pattern Matching Techniques for TGG-Based Model Synchronisation

Graph transformation applications depend highly on the discovery of occurrences of patterns in a host graph (called pattern matching). When an application operates on relatively large models where individual model changes usually concern only a small part, it is impractical to restart the pattern matching process each time from scratch. While auxiliary data structures (such as precedences, look-up tables, or rule application protocols) strive to avoid this in an application-specific manner, incremental pattern matching techniques (e.g., [6]) with recently developed practical solutions (e.g., [17,18]) address the same challenges in a generic and reusable manner. An incremental pattern matcher is capable of maintaining partial and complete matches of a given set of patterns found in a possibly changing host graph. Consequently, appearing or disappearing matches for a given set of patterns can be determined between two points in time (e.g., before and after changing the models). This enables client applications to focus on their business logic and high-level goals by reacting to appearance or disappearance of these matches (without searching and maintaining them manually). We discuss in the following how this vision can be realised for TGG-based model synchronisation.

For forward synchronisation, we propose to monitor two types of patterns in a model triple: *available* and *processed* markings with forward marking rules. In Fig. 4, these patterns are depicted for our running example. Basically, the pattern

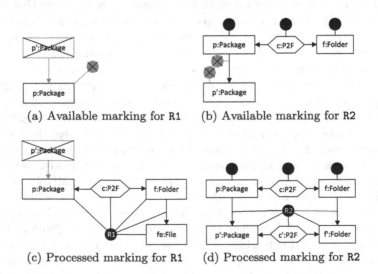

(a) Available marking for R1   (b) Available marking for R2

(c) Processed marking for R1   (d) Processed marking for R2

Fig. 4. Patterns to be monitored by an incremental pattern matcher

for an available marking with a forward marking rule is the precondition (L) of the forward marking rule together with its filter NACs as well as marker NACs. The pattern for a processed marking comprises the postcondition (R) of the forward marking rule together with its filter NACs.

Assuming that matches for available and processed markings are monitored in a model triple by an incremental pattern matcher, Algorithm 1 represents our proposed concept for TGG-based model synchronisation in pseudo code. The procedure PROPAGATESOURCEDELTA takes the following inputs: (i) a consistent model triple $G = G_S \leftarrow G_C \rightarrow G_T$ which is fully marked (from former runs), (ii) a source delta δ_S that changes G_S to G'_S, and (iii) an incremental pattern matcher pm that is initialised with G and monitors patterns for available and processed markings. The outcome of the procedure is a new model triple $G' = G'_S \leftarrow G'_C \rightarrow G'_T$ which reflects the source delta and is again consistent.

Algorithm 1. Model Synchronisation

1: **procedure** PROPAGATESOURCEDELTA(G, δ_S, pm)
2:
3:     $G' \leftarrow$ change G via δ_S ▷ **Phase 1**
4:     pm.update(δ_S)
5:
6:     **while** pm has a disappearing match for processed markings **do**   ▷ **Phase 2**
7:         $m^- \leftarrow$ choose a disappearing match for processed markings
8:         $(G', \delta^-) \leftarrow$ revoke the fwd marking rule for m^- in G'
9:         pm.update(δ^-)
10:    **end while**
11:
12:    **while** pm has an appearing match for available markings **do**   ▷ **Phase 3**
13:        $m^+ \leftarrow$ choose a match for available markings
14:        $(G', \delta^+) \leftarrow$ apply the fwd marking rule for m^+ in G'
15:        pm.update(δ^+)
16:    **end while**
17:
18:    **return** G'
19:
20: **end procedure**

Overall, PROPAGATESOURCEDELTA consists of three main phases:

Phase 1 (Line 3–4): The source delta is applied to the model triple and the incremental pattern matcher updates its matches for available and processed markings in the model triple.

Phase 2 (Line 6–10): Disappearing matches of processed markings indicate that the respective applications of the forward marking rules from former runs are invalidated due to the source delta. Such invalidated rule applications must be revoked by deleting their created markers, correspondences, and target elements.

The incremental pattern matcher updates its matches after these deletions. Note that this can trigger further disappearances of processed marking patterns which again must be handled in the same manner until the pattern matcher does not report any further disappearing match for processed markings.

Phase 3 (Line 12–16): Appearing matches of available markings indicate that the respective forward marking rules are applicable. An arbitrary match is chosen and the forward marking rule is applied by creating a marker, correspondences, and target elements. The incremental pattern matcher updates its matches after these creations. Some matches for available markings disappear (at least the chosen match itself disappears) as some elements are now marked and violate marker NACs. Note that disappearing matches for available markings in this phase indicate progress in the synchronisation process (not to be confused with disappearing matches due to the source delta in Phase 2). Further matches for available markings can also appear due to the creation of new elements and must be handled in the same manner until the pattern matcher does not report any further appearing match for available markings.

In the following, we exemplify the intermediate and end results of Algorithm 1 based on two synchronisation runs with our running example.

Example (initial transformation): We first discuss an initial forward transformation of a source model to a target model. This is a special case of forward synchronisation where the entire source model is a delta applied to an empty triple. We assume that the incremental pattern matcher is initialised with an empty triple and that the source delta is the creation of two outermost packages. Applying this delta in Phase 1, two matches occur as available markings for the forward marking rule of R1, depicted in Fig. 5a via an R1-labeled arrow at the bottom-right corner of each match. No matches for processed markings disappear in this example as the model triple was initially empty, i.e., no rule application is to be revoked in Phase 2. Finally, applying forward marking rules for available markings in Phase 3, two matches occur for processed markings, depicted in Fig. 5b. The model triple is again consistent and fully marked.

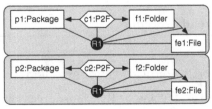

(a) Appearing matches for available markings match after Phase 1 and 2

(b) Consistent state after Phase 3

Fig. 5. Intermediate results of propagating two outermost packages

Example (creating a sub edge): We now assume that the incremental pattern matcher is initialised with the result of the previous example (Fig. 6a) and create a **sub** edge between the two packages making one of them, namely p2, a child package. After applying this delta, a match for a processed marking disappears as p2, being no longer an outermost package, violates the filter NAC. The **sub** edge violating the filter NAC is depicted bold in Fig. 6b while the rectangle with red filling and dashed border represents the disappearing match. Revoking the respective forward marking rule application of this match (i.e., deleting the marker of p2 as well as its corresponding target elements) in Phase 2, a new marking becomes available for the forward marking rule of R2, depicted in Fig. 6c. Applying the forward marking rule for the available marking in Phase 3, the model triple is consistent and fully-marked again (Fig. 6d).

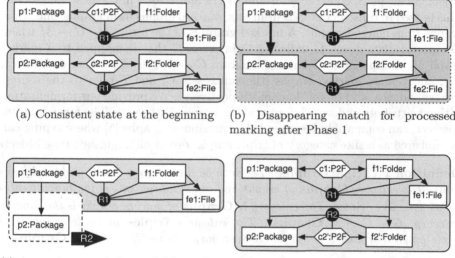

(a) Consistent state at the beginning

(b) Disappearing match for processed marking after Phase 1

(c) Appearing match for available marking after Phase 2

(d) Final consistent state after Phase 3

Fig. 6. Intermediate results of propagating a **sub** edge

Finally, it should be mentioned that the phases of Algorithm 1 represent a straightforward approach without any heuristics to improve the quality of model synchronisation, especially with regard to information preservation capabilities. While Phase 2 revokes rule applications until no more disappearing match is reported, another possible reaction to disappearing matches would be to *repair* them (instead of revoking) as discussed in [7]. Alternatively, target elements deleted in Phase 2 can be *reused* in Phase 3 for new rule applications as proposed in [8]. In both cases, the goal is to preserve as much as possible from the older version of the target model. These extensions are orthogonal to our contribution and can analogously be supported via an incremental pattern matcher. Basically, new types of reactions to appearing/disappearing matches are required for this but the idea of a reactive synchroniser concept remains the same.

4   Correctness of Delta Propagation

We formalise in the following triple graphs and forward marking rules via construction techniques over *functor* categories [5], and prove the correctness of Algorithm 1 under sufficient conditions. Our correctness proof is in line with that of [10,13,15] but eliminates the need of an entire marking from scratch as in [10], or an additional dependency analysis as in [13,15]. The added value of our formalisation thus lies in its simplified form.

A functor category $[\mathcal{S}, \mathcal{C}]$ consists of structure preserving arrows between objects of shape $\mathcal{S}$, constructed from objects and arrows in the host category $\mathcal{C}$. This is used to construct graphs from sets, marked graphs from graphs, and triple graphs from marked graphs. A marked graph is of the form $G \leftarrow \overline{G} \rightarrow M$ where the intermediate graph $\overline{G}$ indicates the part of G that is mapped to a marker graph M. A triple graph is then of the form $G_S \leftarrow G_C \rightarrow G_T$ where each of G_S, G_C, and G_T are marked graphs (the suffixes S, C, and T refer to the source, correspondence, and target domain, respectively). We provide our formalisation without attribute and type information in graphs for brevity. The formalisation, however, can compatibly be extended to attributed graphs [5] where typing can be captured as a slice category of triple graphs over a distinguished type object.

Definition 1 (Triple Graphs). *Let* **Sets** *be the category of sets and total functions. The category* **Graphs** *of graphs and graph morphisms is the functor category* $[\ E\substack{\longrightarrow\\\longrightarrow}V\ ,$ **Sets**$]$. *The category* **MGraphs** *of marked graphs is the functor category* $[G \leftarrow \overline{G} \rightarrow M,$ **Graphs**$]$. *The category* **Triples** *of triple graphs and triple graph morphisms is the functor category* $[G_S \leftarrow G_C \rightarrow G_T,$ **MGraphs**$]$.

Definition 2 (Triple Rule and Derivation). *A triple rule is a morphism* $r : L \rightarrow R$ *in* **Triples**. *A Negative Application Condition (NAC) for a triple rule* $r : L \rightarrow R$ *is a morphism* $n : L \rightarrow N$ *in* **Triples**.

Given a triple rule r *with a set* $\mathcal{N}$ *of NACs, a direct derivation* $G \stackrel{r@m}{\Longrightarrow} G'$ *(or just* $G \stackrel{r}{\Longrightarrow} G'$*) is given by the pushout* $(r' : G \rightarrow G', m' : R \rightarrow G')$ *of* $r : L \rightarrow R$, *and* $m : L \rightarrow G$ *in* **Triples** *if* $\nexists n : L \rightarrow N \in \mathcal{N}, \exists n' : N \rightarrow G, m = n' \circ n$.

A derivation $G \stackrel{*}{\Longrightarrow} G'$ *with a set* $\mathcal{R}$ *of triple rules is a sequence of k direct derivations* $G \stackrel{r_1}{\Longrightarrow} G_1 \stackrel{r_2}{\Longrightarrow} \cdots \stackrel{r_k}{\Longrightarrow} G'$, $r_1, r_2, \cdots, r_k \in \mathcal{R}$ *(*$G' = G$ *for k = 0).*

In a TGG, the original rules do not create or require markers (e.g., Fig. 2) but forward marking rules do (e.g., Fig. 3).

Definition 3 (Triple Graph Grammar).
A triple graph grammar (TGG) is a finite set $\mathcal{R}$ of triple rules, each of the form depicted to the right. A TGG is source progressive if $r_s \neq id$. The language of a TGG is given by $\mathcal{L}(TGG) :=$ $\{G \mid \exists\ G_\emptyset \stackrel{*}{\Longrightarrow} G$ with $\mathcal{R}\}$ where $G_\emptyset$ is the empty triple graph.

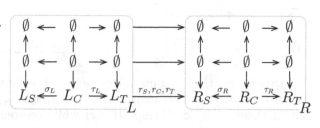

A forward marking rule fr for a TGG rule r as introduced in Definition 3 (i) does not create the elements in $R_S \backslash L_S$ but requires them, (ii) creates all elements in $R_C \backslash L_C$ and $R_T \backslash L_T$, (iii) requires markers for all elements in L_S, L_C, and L_T, (iv) forbids markers for elements in $R_S \backslash L_S$, and finally (v) creates new markers for all elements in $R_S \backslash L_S$, $R_C \backslash L_C$, and $R_T \backslash L_T$ (cf. Fig. 3).

Definition 4 (Forward Marking Rule). *Given a TGG $= \mathcal{R}$, the forward marking rule $fr : FL \rightarrow FR$ for each $r \in \mathcal{R}$ has the structure as follows:*

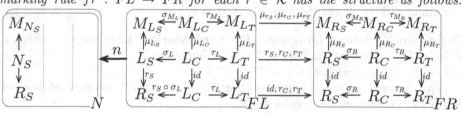

For $X \in \{S, C, T\}$, the graphs M_{L_X} are isomorphic to L_X. The graphs M_{R_X} extend M_{L_X} by an extra node m. All nodes in $R_X \backslash L_X$ are mapped by μ_{R_X} to m. Every edge in $R_X \backslash L_X$ is mapped to an edge added to M_{R_X} so that μ_{R_X} is structure preserving. Furthermore, the forward rule fr is equipped with a set $\mathcal{N}$ of marker NACs. Marker NACs $n : FL \rightarrow N$ extend the source components of FL (all other components remain the same and are not depicted explicitly) to forbid the presence of markers for any element in $R_S \backslash L_S$.

Example: The diagram below depicts the forward marking rule for R1 (Fig. 3a) formally. In FL a node p is required in the source component and the presence of markers for p is forbidden by the marker NAC N. In FR, the correspondence and target elements (c, f, fe, and an edge between f and fe) are created together with a marker m in each component. Note that all nodes that are created in the original TGG rule ($R_X \backslash L_X$) are mapped to one marker m in each component (shown explicitly via dashed lines for f and fe) where edges are mapped to a self-edge of markers (e.g., the edge between f and fe).

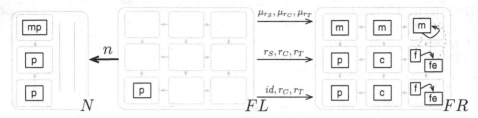

Furthermore, the diagram below depicts the forward marking rule for R2 (Fig. 3b) formally. In this case, markers (mp, mc, mf) are required in FL for context elements (p, c, f, respectively) while created elements as well as created/forbidden markers are analogous to the previous diagram.

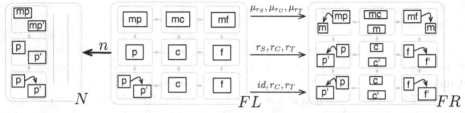

Next, we define the source language of a TGG, i.e., source graphs G_S for which a triple $G_S \leftarrow G_C \rightarrow G_T$ exists in $\mathcal{L}(TGG)$. Accordingly, *the forward marked language* of a TGG is given by derivations via forward marking rules beginning with triples of the form $G_S \leftarrow \emptyset \rightarrow \emptyset$.

Definition 5 (Source Language, Forward Marked Language). *Given a TGG = $\mathcal{R}$, the source language of TGG is defined as*
$\mathcal{L}(TGG)_S = \{G_S \mid \exists\, G_S \leftarrow G_C \rightarrow G_T \in \mathcal{L}(TGG)\}.$
The forward marking grammar fwd(TGG) for TGG consists of the set fwd($\mathcal{R}$) of forward marking rules for $\mathcal{R}$.
The forward marked language of fwd(TGG) for $G_S \in \mathcal{L}(TGG)_S$ is defined as
$\mathcal{L}(fwd(TGG), G_S) = \{G \mid \exists\, (G_S \leftarrow \emptyset \rightarrow \emptyset) \xRightarrow{*} G \text{ with fwd}(\mathcal{R})\}.$

A triple graph is fully marked if every node/edge in its source, correspondence, and target components are mapped to a marker node/edge. Fully marked triples are of interest as every derivation with triple rules from a TGG can be traced back to a unique derivation with the respective forward marking rules where the result is fully marked. We furthermore introduce an operator Φ to extract an unmarked triple graph from a marked one.

Definition 6 (Fully Marked and Unmarked Triple Graphs). *Let $G = G_S \leftarrow G_C \rightarrow G_T$ be a triple graph. G is fully marked if each of its marked graphs G_X, $X \in \{S, C, T\}$, is of the form $X \xleftarrow{id} X \rightarrow M_X$. A triple graph is unmarked if each of its marked graphs is of the form $X \leftarrow \emptyset \rightarrow \emptyset$ and $\Phi(G)$ denotes the unmarked triple graph obtained from G by removing its markers.*

Fact 1 (Bijection Between TGG and Forward Marking Grammar). *Given a TGG with a forward marking grammar fwd(TGG), $\exists\, G = G_S \leftarrow G_C \rightarrow$*

$G_T \in \mathcal{L}(TGG) \iff \exists \; \overline{G} \in \mathcal{L}(fwd(TGG), G_S)$ *where* $\overline{G}$ *is fully marked and* $G = \Phi(\overline{G})$.

Proof (Sketch). This is a standard operationalisation result for TGGs [10], applied to marked graphs and marking rules. Basically, a forward marking rule fr for a TGG rule r maps exactly those source elements to a marker that are created by r. Both r and fr create the same correspondence and target elements whereas fr maps each of them additionally to markers (cf. Definitions 3 and 4). □

We now introduce the notion of *confluence*, a well-known property of graph grammars that can be checked statically [5]. Every partial derivation of a confluent graph grammar can be completed to a derivation that produces the same result, and this ensures (together with Fact 1) that applying forward marking rules (e.g., as in Algorithm 1) results always in a fully marked triple graph.

Definition 7 (Confluence). *A pair* $G_1 \overset{*}{\Longleftarrow} G \overset{*}{\Longrightarrow} G_2$ *of derivations with a set* $\mathcal{R}$ *of triple rules is* confluent *if there exists a* G' *together with derivations* $G_1 \overset{*}{\Longrightarrow} G'$ *and* $G_2 \overset{*}{\Longrightarrow} G'$.
We refer to $\mathcal{R}$ *as* confluent *if all pairs of its derivations are confluent.*

Fact 2 (Confluence of Forward Marking Rules). *Given a* $TGG = \mathcal{R}$ *with forward marking rules* $fwd(\mathcal{R})$, *source language* $\mathcal{L}(TGG)_S$ *and* $G_S \in \mathcal{L}(TGG)_S$, *if* $fwd(\mathcal{R})$ *is confluent, then every derivation* $\tilde{d} = G_S \leftarrow \emptyset \rightarrow \emptyset \overset{*}{\Longrightarrow} \tilde{G}$ *with* $fwd(\mathcal{R})$ *can be extended to a derivation* $\overline{d} = G_S \leftarrow \emptyset \rightarrow \emptyset \overset{*}{\Longrightarrow} \tilde{G} \overset{*}{\Longrightarrow} \overline{G}$, *where* $\overline{G}$ *is fully marked.*

Proof. $G_S \in \mathcal{L}(TGG)_S \overset{Definition\,5}{\Longrightarrow} \exists G_S \leftarrow G_C \rightarrow G_T \in \mathcal{L}(TGG) \overset{Fact\,1}{\Longrightarrow} \exists$ fully marked $\overline{G} \in \mathcal{L}(fwd(TGG), G_S)$. Extension of $\tilde{d}$ to $\overline{d}$ follows from Definition 7 with $G = G_S \leftarrow \emptyset \rightarrow \emptyset, G_1 = \tilde{G}, G_2 = G' = \overline{G}$. □

To ensure practical applicability of TGGs, forward marking rules are often enriched with *filter NACs* (cf., Fig. 3). The goal of filter NACs is to make a non-confluent forward marking grammar confluent [9]. That is, filter NACs block only those derivations that do not lead to a fully marked graph.

Definition 8 (Filter NACs). *Given a* $TGG = \mathcal{R}$ *and its forward marking rules* $fwd(\mathcal{R})$, *each forward marking rule* $fr : FL \rightarrow FR$ *in* $fwd(\mathcal{R})$ *can be equipped with a set* $\mathcal{N}'$ *of filter NACs that do not block any derivation* $G_S \leftarrow \emptyset \rightarrow \emptyset \overset{*}{\Longrightarrow} \overline{G}$ *in* $fwd(\mathcal{R})$ *without filter NACs where* $\overline{G}$ *is fully marked. Filter NACs are, therefore, only used to ensure that* $fwd(\mathcal{R})$ *is confluent. We refer to, e.g., [9] for a construction technique.*

Next, we define matches that are to be monitored in a forward synchronisation process as discussed in Sect. 3 for Algorithm 1. Matches for available markings are simply matches for direct derivations via forward marking rules, while matches for processed markings appear after a direct derivation via forward marking rules and forbid the violation of filter NACs.

Definition 9 (Matches for Available and Processed Markings). *Let* $fr : FL \to FR$ *be a forward marking rule and* $\mathcal{N}'$ *the set of filter NACs of* fr. *Given a triple graph* G, *for each possible direct derivation* $G \overset{fr@m}{\Longrightarrow} G'$, *we refer to* m *as a* match for available marking *with* fr. *A* match for processed marking *with* fr *is given by the comatch* $m' : FR \to G'$. *A match* m' *for processed marking is valid if it does not violate any filter NAC, i.e.,* $\forall n : FL \to N \in \mathcal{N}', \nexists n' : N \to G$ *such that* $n' \circ n = m' \circ fr$.

Deltas represent changes to graphs (to a source graph in case of a forward synchronisation) and lead to appearance or disappearance of matches for available or processed markings.

Definition 10 (Delta). *A delta is a span of graphs and graph morphisms* $\delta_X = G_X \leftarrow \Delta_X \to G'_X$. *Elements in* $G_X \backslash \Delta_X$ *and* $G'_X \backslash \Delta_X$ *are referred to as* deleted *and* created, *respectively, by* δ_X.

Finally, the following Theorem states the correctness of Algorithm 1, i.e., that the result G' is a fully marked triple where $\Phi(G') \in \mathcal{L}(TGG)$. We require a source progressive TGG (Definition 3) with a confluent forward marking grammar (Definition 7) as sufficient conditions.

Theorem 1 (Correctness of Delta Propagation). *Given a source progressive TGG* $= \mathcal{R}$, *let* $G \in \mathcal{L}(fwd(TGG), G_S)$ *be a fully marked triple, and* $\delta_S = G_S \leftarrow \Delta_S \to G'_S$ *a delta. We assume a pattern matcher pm monitoring matches in* G *for available and processed markings with forward marking rules in* $fwd(\mathcal{R})$. *If* $fwd(\mathcal{R})$ *is confluent and* $G'_S \in \mathcal{L}(TGG)_S$, *the result of* PROPAGATESOURCEDELTAG, δ_S, pm *(Algorithm 1) is a triple graph* G' *such that* $\Phi(G') \in \mathcal{L}(TGG)$, *i.e.,* PROPAGATESOURCEDELTAG, δ_S, pm *is correct.*

Proof. We use in the following the intermediate results of Phase 1, 2, and 3 in PROPAGATESOURCEDELTAG, δ_S, pm for the proof.

Phase 1: $G \in \mathcal{L}(fwd(TGG), G_S)$ is fully marked. Hence, there exists a derivation $d_1 = (G_S \leftarrow \emptyset \to \emptyset) \overset{*}{\Longrightarrow} G$ with $fwd(\mathcal{R})$. The comatch of each direct derivation in d_1 leads to a match for processed marking (Definition 9). When changing G_S according to δ_S, such matches can disappear while new matches for available markings can appear.

Phase 2: For each disappearing match for processed markings, direct derivations with $fwd(\mathcal{R})$ in d_1 are revoked. This step does not create any match for processed markings and thus terminates when no more matches disappear (d_1 has finitely many direct derivations). With the remaining direct derivations from d_1 (i.e., direct derivations that have not been revoked), we get a derivation $d_2 = G'_S \leftarrow \emptyset \to \emptyset \overset{*}{\Longrightarrow} \widetilde{G}$ with $fwd(\mathcal{R})$.

Phase 3: Given that $G'_S \in \mathcal{L}(TGG)_S$ and $fwd(\mathcal{R})$ is confluent, d_2 can be extended to a derivation $d_3 = G'_S \leftarrow \emptyset \to \emptyset \overset{*}{\Longrightarrow} \widetilde{G} \overset{*}{\Longrightarrow} G'$ by applying available marking matches in any order where the result is a fully marked triple G' due to

Fact 2 and $\Phi(G') \in \mathcal{L}(TGG)$ due to Fact 1. Termination of this step is guaranteed as the TGG is source progressive (Definition 3), i.e., each direct derivation marks at least one source element and reduces the number of source elements for which available matches can appear (due to marker NACs in Definition 4). $\square$

5   Related Work

Bx approaches can be classified mainly in three categories: relational (e.g., [3]), programming-based (e.g., [12]), and rule-based approaches such as TGGs. Our contribution exploits the rule-based characteristics of TGGs and governs a synchronisation process via appearing/disappearing rule matches. We believe, nevertheless, that at least relation-based approaches can be inspired by our contribution to monitor (un-)satisfied relations between two models.

We have already discussed in Sect. 2 different TGG approaches [7,8,10,11, 13,15] to emphasise what our contribution exactly simplifies with regard to the runtime tasks of a TGG-based synchronisation. While our focus is the underlying technology of TGG-based synchronisation, practical extensions including repair rules [7] or reusing deleted elements [8] are useful to improve the quality of TGG-based synchronisation (by keeping as many elements as possible from the older versions of models). An incremental pattern matcher can facilitate such extensions by introducing new types of reactions to appearing/disappearing matches. Furthermore, the confluence requirement in our formalisation can be relaxed via static analysis techniques [1] such that model synchronisation can have more than one possible valid result. This is orthogonal to our contribution and non-confluence has not been considered due to space limitations.

Our work is inspired by model synchronisation applications operating with incremental pattern matching techniques. Most closely, Bergmann et al. [2] demonstrate how to transform a source delta to a target delta by using incremental pattern matchers. The transformation step, however, is a manually implemented forward transformation, while TGGs introduce a grammatical and declarative consistency notion and the forward transformation is automatically derived together with its backward counterpart.

6   Conclusion and Future Work

We have presented a novel concept for TGG-based model synchronisation based on an underlying incremental pattern matcher. A TGG-based synchroniser is reduced to a component that simply reacts to appearing or disappearing matches monitored by its underlying incremental pattern matcher. We have formalised our synchroniser concept and shown its correctness under sufficient conditions.

Future work has already started on, first and foremost, implementing a feature-complete TGG tool with our concept and evaluating its capabilities with comparisons to other TGG and bx approaches. Incremental pattern matchers strive for scalable computations of matches whose runtime depends on delta size and not on model size. Our expectation, therefore, is that improved scalability will be a second advantage besides the simplified synchroniser concept when integrating incremental pattern matchers into TGGs. This is yet to be validated by extending the comparison of TGG-based model synchronisation tools [14].

We are also interested in (incremental) consistency checking and model integration (two-way model synchronisation potentially with conflict resolution) with TGGs. Such advanced use cases become tractable and can be handled uniformly after abstracting TGGs from low-level details of match maintenance.

Acknowledgement. This work has been funded by the German Federal Ministry of Education and Research within the Software Campus project GraTraM at TU Darmstadt, funding code 01IS12054.

References

1. Anjorin, A., Leblebici, E., Schürr, A., Taentzer, G.: A static analysis of non-confluent triple graph grammars for efficient model transformation. In: Giese, H., König, B. (eds.) ICGT 2014. LNCS, vol. 8571, pp. 130–145. Springer, Cham (2014). doi:10.1007/978-3-319-09108-2_9
2. Bergmann, G., Ráth, I., Varró, G., Varró, D.: Change-driven model transformations - change (in) the rule to rule the change. SoSym **11**(3), 431–461 (2012)
3. Cicchetti, A., Ruscio, D., Eramo, R., Pierantonio, A.: JTL: a bidirectional and change propagating transformation language. In: Malloy, B., Staab, S., Brand, M. (eds.) SLE 2010. LNCS, vol. 6563, pp. 183–202. Springer, Heidelberg (2011). doi:10.1007/978-3-642-19440-5_11
4. Czarnecki, K., Foster, J.N., Hu, Z., Lämmel, R., Schürr, A., Terwilliger, J.F.: Bidirectional transformations: a cross-discipline perspective. In: Paige, R.F. (ed.) ICMT 2009. LNCS, vol. 5563, pp. 260–283. Springer, Heidelberg (2009). doi:10.1007/978-3-642-02408-5_19
5. Ehrig, H., Ehrig, K., Prange, U., Taentzer, G.: Fundamentals of Algebraic Graph Transformation. Monographs in Theoretical Computer Science. An EATCS Series. Springer, Heidelberg (2006)
6. Forgy, C.: Rete: a fast algorithm for the many patterns/many objects match problem. Artif. Intell. **19**(1), 17–37 (1982)
7. Giese, H., Hildebrandt, S.: Efficient model synchronization of large-scale models. Technical report, HPI at the University of Potsdam (2009)
8. Greenyer, J., Pook, S., Rieke, J.: Preventing information loss in incremental model synchronization by reusing elements. In: France, R.B., Kuester, J.M., Bordbar, B., Paige, R.F. (eds.) ECMFA 2011. LNCS, vol. 6698, pp. 144–159. Springer, Heidelberg (2011). doi:10.1007/978-3-642-21470-7_11
9. Hermann, F., Ehrig, H., Golas, U., Orejas, F.: Efficient analysis and execution of correct and complete model transformations based on triple graph grammars. In: Bézivin, J., Soley, R.M., Vallecillo, A. (eds.) MDI 2010, pp. 22–31. ACM (2010)
10. Hermann, F., Ehrig, H., Orejas, F., Czarnecki, K., Diskin, Z., Xiong, Y.: Correctness of model synchronization based on triple graph grammars. In: Whittle, J., Clark, T., Kühne, T. (eds.) MODELS 2011. LNCS, vol. 6981, pp. 668–682. Springer, Heidelberg (2011). doi:10.1007/978-3-642-24485-8_49
11. Klassen, L., Wagner, R.: EMorF - a tool for model transformations. ECEASST **54** (2012)

12. Ko, H.-S., Zan, T., Hu, Z.: BiGUL: a formally verified core language for putback-based bidirectional programming. In: PPEPM 2016, pp. 61–72. ACM, New York (2016)
13. Lauder, M., Anjorin, A., Varró, G., Schürr, A.: Efficient model synchronization with precedence triple graph grammars. In: Ehrig, H., Engels, G., Kreowski, H.-J., Rozenberg, G. (eds.) ICGT 2012. LNCS, vol. 7562, pp. 401–415. Springer, Heidelberg (2012). doi:10.1007/978-3-642-33654-6_27
14. Leblebici, E., Anjorin, A., Schürr, A., Hildebrandt, S., Rieke, J., Greenyer, J.: A comparison of incremental triple graph grammar tools. ECEASST **67** (2014)
15. Orejas, F., Pino, E.: Correctness of incremental model synchronization with triple graph grammars. In: Ruscio, D., Varró, D. (eds.) ICMT 2014. LNCS, vol. 8568, pp. 74–90. Springer, Cham (2014). doi:10.1007/978-3-319-08789-4_6
16. Schürr, A.: Specification of graph translators with triple graph grammars. In: Mayr, E.W., Schmidt, G., Tinhofer, G. (eds.) WG 1994. LNCS, vol. 903, pp. 151–163. Springer, Heidelberg (1995). doi:10.1007/3-540-59071-4_45
17. Ujhelyi, Z., Bergmann, G., Hegedüs, Á., Horváth, Á., Izsó, B., Ráth, I., Szatmári, Z., Varró, D.: EMF-IncQuery: an integrated development environment for live model queries. Sci. Comput. Program. **98**(1), 80–99 (2015)
18. Varró, G., Deckwerth, F.: A rete network construction algorithm for incremental pattern matching. In: Duddy, K., Kappel, G. (eds.) ICMT 2013. LNCS, vol. 7909, pp. 125–140. Springer, Heidelberg (2013). doi:10.1007/978-3-642-38883-5_13

Henshin: A Usability-Focused Framework for EMF Model Transformation Development

Daniel Strüber[1]([✉]), Kristopher Born[2], Kanwal Daud Gill[1], Raffaela Groner[3], Timo Kehrer[4], Manuel Ohrndorf[5], and Matthias Tichy[3]

[1] Universität Koblenz-Landau, Koblenz, Germany
{strueber,daud}@uni-koblenz.de
[2] Philipps-Universität Marburg, Marburg, Germany
born@mathematik.uni-marburg.de
[3] Universität Ulm, Ulm, Germany
{raffaela.groner,matthias.tichy}@uni-ulm.de
[4] Humboldt-Universität zu Berlin, Berlin, Germany
timo.kehrer@informatik.hu-berlin.de
[5] Universität Siegen, Siegen, Germany
mohrndorf@informatik.uni-siegen.de

Abstract. Improved usability of tools is a fundamental prerequisite for a more widespread industrial adoption of Model-Driven Engineering. We present the current state of Henshin, a model transformation language and framework based on algebraic graph transformations. Our demonstration focuses on Henshin's novel usability-oriented features, specifically: (i) a textual syntax, complementing the existing graphical one by improved support for rapid transformation development, (ii) extended static validation, including checks for correct integration with general-purpose-language code, (iii) advanced refactoring support, in particular, for splitting large transformation programs, (iv) editing utilities for facilitating recurring tasks in model transformation development. We demonstrate the usefulness of these features using a running example.

1 Introduction

Model-Driven Engineering (MDE) aims to improve the productivity of software engineers by emphasizing model transformation as a central activity during software development [1]. Still, a major roadblock to a more widespread adoption of MDE is the insufficient maturity of MDE tools [2,3]. Specifically, to make MDE tools appealing to a broader user base, it is key to increase their level of usability.

Henshin [4] is a model transformation framework for the Eclipse Modeling Framework, comprising a transformation language with a graph-transformation-based visual syntax, and a tool environment with an execution engine and analysis features, including model checking and conflict analysis support. Originally designed to offer the benefits of a solid formal foundation and efficient transformation execution, Henshin was not built with usability as an explicit goal.

In fact, based on user feedback, we identify a number of critical usability limitations: First, while its visual syntax is beneficial when *reading* a transformation

J. de Lara and D. Plump (Eds.): ICGT 2017, LNCS 10373, pp. 196–208, 2017.
DOI: 10.1007/978-3-319-61470-0_12

program, *writing* a transformation program can be complicated due to layouting issues. Second, programs can contain subtle errors that are not caught by adequate static checks. In particular, this applies when transformations are not specified in isolation, but embedded into a richer software infrastructure. Third, when working with large transformation programs, scalability issues occur; the performance of Henshin's visual editor may suffer to the point that it becomes unusable [5]. Fourth, users are required to perform intricate and error-prone tasks, such as creating large rules that reflect the complexity of the involved meta-models.

Therefore, in Sect. 2 of this paper, we present the current state of Henshin, focusing on its novel features for addressing these issues. Specifically,

- we introduce a **textual syntax** for the rapid development of transformations (Sect. 2.1). This syntax is not intended as a replacement for the graphical one, but as a complementary means to facilitate the initial creation of a transformation program. To support long-term maintenance, we provide a higher-order transformation that can be used to derive a graphical concrete-syntax representation of the transformation program. The design of our textual syntax was informed by a qualitative interview study.
- we provide extended **static checks** for validating the well-definedness of a transformation and its use (Sect. 2.2). Using this checks, one can validate if a Henshin transformation program is used correctly in general-purpose-language code, e.g., if all referred rules actually exist and their parameters are assigned correctly. Furthermore, we provide checks to see if parameters are specified and used correctly within and across particular units and rules.
- we present advanced **refactoring** support, in particular, for splitting a large transformation program into multiple sub-programs (Sect. 2.3). Using a wizard, the user can specify target sub-programs and assign particular units and rules to them. This splitting of programs (i.e. the abstract syntax) can also be propagated to their diagram files (i.e., the concrete syntax).
- we demonstrate a selection of **editing utilities** for complicated tasks during the development of transformations (Sect. 2.4), including utilities to create, simplify, generalize, or clean up Henshin rules.

Running Example. We use the following transformation program as a running example throughout this paper. The program solves one of the tasks in the classical Comb benchmark by Varró et al. [6]: It constructs a *sparse grid* in the shape of Fig. 1 for a given pair of dimensions, *width* and *height*. Note that the grid has two kinds of edges: vertical and horizontal ones (dashed and bold arrows, respectively).

In Henshin, programs are specified in the form of *modules*. A module contains a set of *rules*, specifying in-place transformations, and a set of composite *units*, managing the control flow. Specifically, composite units coordinate the execution of their sub-units, which can be either rules or other composite units.

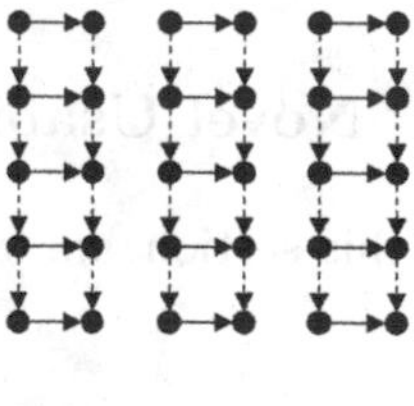

Fig. 1. Sparse grid

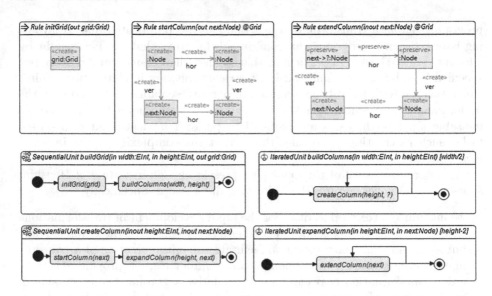

Fig. 2. Henshin module (program) for building a sparse grid.

The example module shown in Fig. 2 includes three rules and four units. The entry point is the sequential unit *buildGrid* which has two input parameters, *width* and *height*, and one output parameter, *grid*. This unit calls two sub-units in sequential order: rule *initGrid* creates an initially empty grid to be delivered as output of the overall transformation. Iterated unit *buildColumns* has an iteration condition, specifying that its sub-unit *createColumn* is executed *width*/2 times. Sequential unit *createColumn* uses parameter *next* as a pointer to build a column of a particular height; *next* is initialized in rule *startColumn*, where the first two rows of a column are created. Note the syntactic sugar @Grid, which specifies the presence of a grid to be used as a container for all newly created nodes. Parameters *next* and *height* are passed to iterated unit *expandColumn*, which executes rule *extendColumn height-2* times. Each execution of *extendColumn* uses the *next* pointer to add another column at this node, changing the pointer to one of the newly created nodes afterwards. The program terminates after all columns have been constructed by unit *buildColumns*, yielding the sparse grid.

This graphical syntax uses a compact representation of rules, where left- and right-hand sides (LHS, RHS) are combined to one graph with annotations such as «create» for RHS nodes without a LHS counterpart. The abstract syntax maintained in the background of the visual editor captures LHS and RHS explicitly.

2   Novel Usability-Oriented Features

In this section, we walk through the novel usability-oriented features.

2.1   Textual Syntax for Henshin

While much of research and practice in modeling has focused on graphical modeling languages, there has been a trend in recent years to use textual concrete syntaxes for modeling languages. The reason for that is the increasing feedback from industrial practice that graphical editors (1) require a high effort to create something usable in practice (which has been shown as a huge issue in modeling in practice [3]), (2) require syntax correctness, resulting in cumbersome user actions to avoid intermediate incorrect models during complex editing steps, and (3) lack acceptance by many end-users, particularly, software developers.

The resurgence of modeling languages with a textual syntax is fueled by easy-to-use and feature-rich frameworks like Xtext and recent advances in projection-based modeling frameworks [7]. Finally, there is a trend to seamlessly combine graphical and textual editors to reap benefits of both worlds [8].

Henshin currently supports both a graphical editor as well as a tree-based editor. The graphical editor supports some syntactic sugar in both syntax and visualization, e.g., combined notation for LHS and RHS, NACs, container syntax. However, one needs to use the tree-based editor for more complex rules, e.g., rules with complex nesting of condition graphs. Furthermore, the usability in terms of efficiency

```
1 rule initGridWithTwoNodes(OUT
      grid:Grid){
2   graph{
3     create node one:Node
4     create node two:Node
5     create node grid:Grid
6     edges[(create one->two:ver),
7        (create grid->one:nodes),
8        (create grid->two:nodes)
9     ]
10  }
11 }
```

Fig. 3. Simple rule in the textual syntax

is quite low as both the graphical editor and the tree-based editor require many steps to perform changes and high amount of focus change when moving between the graphical editor pane and the property editor.

Methodology. Last year, we performed a qualitative study to explore different alternatives of a potential textual syntax for Henshin. Particularly, we focused on the following variation points:

combined syntax vs. explicit LHS and RHS. Shall a combined syntax using mark-ups for created and deleted elements and positive and negative applications conditions be used as in the graphical editor, or a specific LHS and RHS as in the tree-based editor?

complex application conditions. How shall complex application conditions with multiple condition graphs be specified?

control flow specification. Shall the control flow specification follow the current Henshin style of different units for different control flow constructs, e.g., sequences, conditions, loops, priorities or should the textual syntax resemble typical programming languages?

```
 1 rule startNextColumn(){
 2    graph{
 3       node root:Grid
 4       node unnamed:Node
 5       create node newNode:Node
 6       edges[(root->unnamed:nodes),
 7          (create root->newNode:nodes),
 8          (create unnamed->newNode:hor)]
 9       matchingFormula{
10          formula !graph1 AND !graph2
11          conditionGraph graph1{
12             node forbidNode:Node
13             edges[(root->forbidNode:nodes),
14                (forbidNode->unnamed:ver)]
15          }
16          conditionGraph graph2{
17             node forbidNode:Node
18             edges[(root->forbidNode:nodes),
19                (unnamed->forbidNode:hor),
20                (root->unnamed:nodes)]
21          }
22       }
23    }
24 }
```

Fig. 4. Rule with complex conditions

Furthermore, we explored other syntax variation points such as syntax for the specification of nodes, edges, and attribute assignments.

We built multiple prototypes covering the different variation points and discussed them in an interview study with 6 current and former Henshin key developers covering a diverse set of expertise in language design and experience using and developing Henshin. The interviews were based on a semi-structured questionnaire, covering demographics on the interviewees, general questions on textual vs. graphical editors and the mentioned variation points of the prototypes. They took between 1 and 1.5 h and were executed by two of the authors of this paper. The interviews were transcribed and analyzed using thematic analysis.

Threats to Validity. The external validity of our methodology is threatened by the fact that we only interviewed advanced users. Arguably, advanced users can particularly benefit from a textual syntax since they write more complicated programs, in which the limitations of graphical syntax are more obvious. Still, it yet needs to be studied if our design decisions are also useful for novice users.

Language Design. The general conclusion with respect to the first variation point was that a combined syntax using markups as in the graphical editor and in various other graph transformation tools is preferable. Figure 3 shows a transformation rule from "Full Grid", a slightly modified version of our running example.

Figure 4 shows a transformation rule with complex condition graphs. The example describes the creation of a new initial node in a new column connected to the top node of an existing column, i.e., it neither has a horizontal incoming node nor a vertical outgoing node. Complex conditions are defined with multiple conditionGraphs and the formula keyword which contains the complex boolean condition. Furthermore, the example contains the implicit reuse of nodes of the LHS in the conditionGraphs on the example of the node unnamed. This is similarly possible for multi rules.

Finally, Fig. 5 shows the addColumns transformation unit. In contrast to standard Henshin where for each type of control flow (sequence, if, loop) covering multiple rules an individual unit has to be declared, the textual syntax provides syntax constructs which are similar to imperative programming languages.

```
1 unit addColumns
2   (IN width:EInt, IN height:EInt){
3     for(width - 1){
4       startNextColumn()
5       expandNextColumn(height)
6     }
7 }
```

Fig. 5. Units

Realization. We realized the textual syntax editor using Xtext with custom extensions like scoping and syntax validation. Since the language differs in syntax significantly from the Henshin meta-model, we used the Xtext generated meta-model. Instances of that meta-model are transformed by model transformations to instances of the standard Henshin meta-model. Doing so, we can reuse Henshin's visual syntax, and its interpretation and analysis plug-ins. The realized plug-ins contain automated test for the generated parser as well as automated tests for the transformation.

2.2 Static Checks

Identification and fixing of errors in the development process of a transformation program as early as possible is crucial for user satisfaction. Static checks provide an important feedback to identify such errors. Henshin supports three groups of static checks: (1) basic checks regarding the well-formedness of units and rules, (2) semantic checks regarding consistency preservation and potential mismatches between intended and specified meaning, and (3) checks for the validity of code for loading and executing units and rules. Identified violations are reported to the user using Eclipse's warning and errors markers in the respective editors.

Well-Definedness Checks. Violations to well-definedness constraints are detected and highlighted with an error marker. First, this applies to obvious issues such as rules and units with duplicate signatures, i.e., identical name and parameter lists. Second, parameter handling inside and between rules is checked.

Parameters have a name, type, and kind, where the kinds *in, out, inout, var* specify the usage context: *in* and *inout* parameters have to be set externally; *var* and *out* parameters are set during the rule application. The value of *var* parameters is hidden to the outside world, whereas *in, out*, and *inout* parameters can be used to pass values between rules and units. Checks ensure that parameters are used consistently to their kind (e.g., a *var* parameter must be used in the LHS of a rule) and that parameters are passed consistently. For example, *inout* parameter *next* of rule *extendColumn* requires all units invoking the rule to specify a value, which is the case since unit *expandColumn* passes its own *next* parameter value.

Semantic checks. With semantic checks, we catch some mistakes that are frequently made by novice users. For example, if a node is to be deleted, double-pushout semantics requires that the deletion of all adjacent edges—in the case of EMF, at least the containment edge—is specified as well. Users unaware of this fact might be puzzled when an affected rule cannot be applied. Thus, for delete nodes specified without a containment edge, we show a warning (rather than an error, to support corner cases where a single root node is deleted). Moreover, we provide checks to identify rules that threaten model consistency (see [9]). For example, the application of a rule may not create containment cycles.

Integration with Java Code. Transformation programs are often used in the context of larger programs, such as Eclipse plug-ins. Henshin's Java API provides an interface for loading transformation programs, applying them, and saving the results. The API requires that

```
66 ruleApp = new RuleApplicationImpl(engine);
67 ruleApp.setUnit(module.getUnit("init_grid"));
68
69 unitApp = new UnitApplicationImpl(engine);
70 unitApp.setUnit(module.getUnit("buildGrid"));
71 unitApp.setParameterValue("WIDTH", 2);
72 unitApp.setParameterValue("height", "two");
```

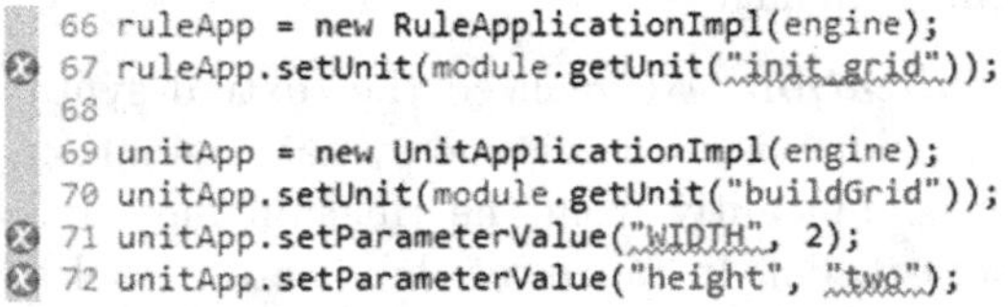

Fig. 6. Error markers in the programmatic use.

certain inputs, such as unit and rule names and parameter values, are provided using method parameters. Errors such as mismatches between specified and allowed values can occur that are only discovered at runtime – a drawback resulting from Henshin's interpreter semantics. To mitigate this drawback while keeping the benefits of interpreted languages, such as flexibility w.r.t. higher-order transformations, we introduce custom checks. In Fig. 6, typos `init_grid` and `WIDTH` are detected since no suitable elements of these names exist in the input module; `"two"` yields a type error as an `int` was expected.

2.3   Advanced Refactoring

Refactorings aim to improve the non-functional properties of a program, without changing its behavior. We distinguish *advanced refactorings* for achieving a higher-level goal from fine-grained *micro-refactorings*. In Henshin, typical micro-refactorings such as *rename rule* and *move rule* are supported by design: Henshin inherits EMF's features for changing models consistently in the sense that references to renamed or moved elements within the same model remain valid automatically. Therefore, in this section, we focus on advanced refactorings.

Transformation programs are typically developed iteratively. The user starts with a small module that easily fits into one screen, such as the one in Fig. 2, and end up with a large module with dozens of rules. Maintaining such a large module is difficult: Navigating the resulting diagram becomes tedious quickly; the performance of the editor may suffer to the point that it is not usable anymore.

Splitting of Modules. To overcome these limitations, we provide a split refactoring that takes a module, and partitions the contained rules and units into sets that are saved into distinct modules. The splitting works in two steps: First, using the wizard shown in Fig. 7, a splitting specification is created. Users can edit the specification by adding and removing target modules, and by reassigning rules and units using drag and drop functionality. In this example, two separate target modules are specified, one including all rules, the other including all units.

As an aid to reduce the manual specification effort, the button "Groups" produces a default suggestion based on an connected-component analysis of the call graph of units, so that each component of units and rules becomes a module. Second, the modules are created and populated with the specified rules and units. The splitting of the concrete syntax models is propagated to the diagram files, i.e., for each target module, a corresponding pair of model and diagram files is created.

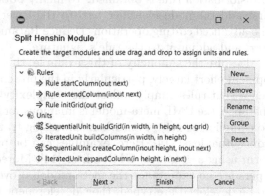

Fig. 7. Splitting wizard.

Merging of Rules. A special type of complexity in transformations arises when many similar rules are required to achieve a common task. In earlier work, we extended Henshin with mechanism detecting rule clones [10] and merging them into an integrated representation that users can interact with [11,12].

2.4   Advanced Editing Support

In this section, we introduce advanced editing utilities for the development of model transformations. First, we present a basic utility that enables users to infer an initial version of a transformation rule from existing models instead of creating the rule from scratch. Second, we give a set of complex editing operations for simplifying, generalizing and cleaning up transformation programs.

Generation of Transformation Rules. We provide a facility to generate a transformation rule from a pair of models demonstrating the effect of the rule, following the principle of model transformation by-example [13]. Technically, the models serving as input of our rule generation procedure are compared with each other using EMF Compare in order to identify the corresponding elements in the original and the changed model, i.e. those elements which are considered to be the same in both models. Thereupon, a transformation rule is basically generated as follows: The original model is converted to a Henshin graph and used as the LHS of the resulting rule, while the Henshin graph obtained from the changed model is used as the RHS. Finally, LHS-RHS mappings are created for all nodes obtained from a pair of corresponding elements.

For instance, an initial version of the rule *startColumn* shown in Fig. 2 can be obtained from an example where the original model contains a container element of type *Grid* and the changed model includes a *Grid* element including the four elements of type *Node* and their respective connections. Thereupon, a Henshin module including the necessary meta-model imports and the generated transformation rule is obtained. To finally obtain the rule *startColumn*, we may first add the node identifier *next* to the lower left node, and subsequently use the advanced editing operation "Deduce Parameters" creating the out parameter *next* and adding it to the rule's signature.

Despite the simplicity of this example, this mechanism can reduce the development effort largely, particularly in the presence of accidental complexity in the sense that rules simply reflect the complexity of the involved meta-model. For instance, the UML meta-model is infamous for its size and complexity that leads to complicated rules even when expressing transformations that are simple on a conceptual level [14]. Here, examples can be provided in a much more compact form using a graphical UML editor [14]. Moreover, when transformation developers are no experts for the meta-model(s) over which the transformation rules to be developed are typed, their development from scratch is likely to be error-prone, e.g. because developers forget to specify certain edges which may lead to unexpected violations of dangling reference constraints in case of deletions.

Complex Editing Operations. In the sequel, we present a set of complex editing operations, each of them taking a structural element of an existing transformation (*Module, Unit* or *Rule*) as input and performing the update of the given element in an in-place fashion. All operations are made available through the "Advanced Editing" menu of the Henshin editor, their visibility depends on the selected context element. Each of these operations formalizes a recurring task during transformation development.

reduceToMinimalRule(Rule) takes a transformation rule as input and reduces it to the minimal rule yielding the same effect. Essentially, it (i) deletes all application conditions and (ii) cuts all context not needed to achieve the specified change, i.e. elements to be preserved by the rule which are not serving as boundary element of a change action are being deleted.

generalizeNodeTypes(Rule) takes a transformation rule as input and converts the types of all LHS nodes to the most general yet still valid supertype. Given a LHS node n of type T, then a supertype T_{sup} of T is a valid supertype in this context if all edges incident to n may be also incident to nodes of type T_{sup} without violating the conformance to the underlying meta-model.

cleanUp(Rule) has been introduced in an earlier Henshin release, e.g. to remove invalid LHS-RHS mappings as well as invalid multi-mappings from the given input rule. We extend this editing operation by also deleting unused rule parameters, i.e., all rule parameters which are neither mapped to a node identifier nor to an attribute variable.

3 Related Work

Usability in Model Transformations. Most works on addressing usability during model transformations focus on the usability in the resulting system. Panach et al. [15] propose to map reusable usability patterns, such as *cancel*, *undo*, and *warnings*, to system models and their transformations, so that the generated systems can benefit from these features. Ammar et al. [16] investigate parametrized model transformations, where usability requirements can be used to select the most suitable among different alternatives. The paradigm of end-user model transformation [14] enables users to specify model transformations using regular model editors. This approach is orthogonal to ours, as it aims to replace specialized transformation editors, rather than to improve them.

Textual Syntax. Interestingly, most model transformation languages based on the graph transformation formalism provide a graphical concrete syntax, e.g., Fujaba, VMTS, ModGraph, whereas other model transformation languages like QVT, ATL, provide a textual concrete syntax. Viatra2 [17] and GrGen [18] are the exceptions as they are based on graph transformations but provide a textual syntax. PROGRES [19] uses an hybrid syntax, whereas eMoflon [20] offers a textual syntax with a generated read-only visualisation, an interesting compromise for supporting both textual and visual notations. Arguably, in our case, having an editable graphical syntax is beneficial since users may use custom layout to give cues beyond the formal language semantics. To the best of our knowledge our work is the first empirical work on designing a textual syntax for model transformations.

Static Checks. Existing works on verifying or testing model transformations focus on correctness w.r.t. a behaviour specification [21,22]. In contrast, our semantic checks can be applied when no specification is available: they represent an heuristics based on accrued experiences with users who were unfamiliar with Henshin's double-pushout semantics. Moreover, to the best of our knowledge, the validation of code that uses a transformation has not been considered before. It is worth pointing out that such validation only make sense for interpreted languages such as Henshin. Compiled ones such as PROGRES avoid the encountered issues via the type system of the target language. However, compilation has certain trade-offs, such as less flexible usage workflows. PROGRES also allows defining precisely how external code is to be integrated with the system.

Advanced Refactoring. Our *split module* refactoring is inspired by an earlier tool-supported approach to meta-model splitting [23]. While this earlier work used clustering algorithms to identify groups of related classes, our default splitting suggestion is based on a component analysis of the call graph of units and rules. Another related work allows remodularizing ATL transformations using clustering [24], where the focus was on the identification of explicit interfaces.

Editing Utilities. Our most advanced editing utility is the generation of rules from existing examples. Learning model transformations from examples is highly desirable and has motivated a plethora of work, as surveyed in [13,25]. However, most approaches are not usable for our purposes since they (i) rely on the logging of editing commands demonstrating the transformation, or (ii) target model-to-model transformations. For state-based approaches targeting in-place transformations, tool support is scarcely available and, to the best of our knowledge, none of the existing tools generates Henshin rules. Our current solution is lightweight in the sense that it abstains from using sophisticated inference algorithms, machine learning techniques or other third-party software, which was a major design decision to keep the deployment of Henshin easy to handle.

4   Conclusion and Future Work

Compared to purely textual languages, where developers have to piece together graph structures in their minds while reading a model transformation program, it is tempting to view graph-based ones as inherently user-friendly. Still, experience has shown that the devil is often in the details: while particular usability issues might not be obvious in smaller examples, the development of larger transformation program can be a tedious task. With the present work, we make a number of contributions to resolve the issues encountered during such tasks.

In the future, we are interested to study the impact of our usability-oriented features. An empirical user study would be an appropriate basis to determine the usefulness of our contributions. A key challenge for advanced features such as those introduced here is to make users aware of usage opportunities [26]. Furthermore, we plan to extend Henshin with additional features. Inferring transformation rules from a set of examples instead of a single one is an interesting goal, in which we may benefit from the existing literature on transformation-by-example. Finally, in our ongoing work, we aim to support the debugging of Henshin rules via an integration into the Eclipse Debugging infrastructure.

Acknowledgement. We thank the reviewers for their valuable and constructive suggestions. This research was partially supported by the research project Visual Privacy Management in User Centric Open Environments (supported by the EU's Horizon 2020 programme, Proposal number: 653642). This work was partially supported by the DFG (German Research Foundation) (grant numbers TI 803/2-2 and TI 803/4-1).

References

1. Sendall, S., Kozaczynski, W.: Model transformation: the heart and soul of model-driven software development. IEEE Softw. **20**(5), 42–45 (2003)
2. Whittle, J., Hutchinson, J., Rouncefield, M., Burden, H., Heldal, R.: Industrial adoption of model-driven engineering: are the tools really the problem? In: Moreira, A., Schätz, B., Gray, J., Vallecillo, A., Clarke, P. (eds.) MODELS 2013. LNCS, vol. 8107, pp. 1–17. Springer, Heidelberg (2013). doi:10.1007/978-3-642-41533-3_1
3. Liebel, G., Marko, N., Tichy, M., Leitner, A., Hansson, J.: Assessing the state-of-practice of model-based engineering in the embedded systems domain. In: Dingel, J., Schulte, W., Ramos, I., Abrahão, S., Insfran, E. (eds.) MODELS 2014. LNCS, vol. 8767, pp. 166–182. Springer, Cham (2014). doi:10.1007/978-3-319-11653-2_11
4. Arendt, T., Biermann, E., Jurack, S., Krause, C., Taentzer, G.: Henshin: advanced concepts and tools for in-place EMF model transformations. In: Petriu, D.C., Rouquette, N., Haugen, Ø. (eds.) MODELS 2010. LNCS, vol. 6394, pp. 121–135. Springer, Heidelberg (2010). doi:10.1007/978-3-642-16145-2_9
5. Strüber, D., Kehrer, T., Arendt, T., Pietsch, C., Reuling, D.: Scalability of model transformations: position paper and benchmark set. In: Workshop on Scalable Model Driven Engineering (BigMDE), pp. 21–30 (2016)
6. Varró, G., Schurr, A., Varró, D.: Benchmarking for graph transformation. In: Symposion on Visual Languages and Human-Centric Computing, pp. 79–88. IEEE (2005)
7. Voelter, M., Szabó, T., Lisson, S., Kolb, B., Erdweg, S., Berger, T.: Efficient development of consistent projectional editors using grammar cells. In: International Conference on Software Language Engineering (SLE), pp. 28–40 (2016)
8. Maro, S., Steghöfer, J., Anjorin, A., Tichy, M., Gelin, L.: On integrating graphical and textual editors for a UML profile based domain specific language: an industrial experience. In: International Conference on Software Language Engineering (SLE), pp. 1–12 (2015)
9. Biermann, E., Ermel, C., Taentzer, G.: Formal foundation of consistent EMF model transformations by algebraic graph transformation. Softw. Syst. Model. **11**(2), 227–250 (2012)
10. Strüber, D., Plöger, J., Acreţoaie, V.: Clone detection for graph-based model transformation languages. In: Van Gorp, P., Engels, G. (eds.) ICMT 2016. LNCS, vol. 9765, pp. 191–206. Springer, Cham (2016). doi:10.1007/978-3-319-42064-6_13
11. Strüber, D., Rubin, J., Arendt, T., Chechik, M., Taentzer, G., Plöger, J.: *Rule-Merger*: automatic construction of variability-based model transformation rules. In: Stevens, P., Wąsowski, A. (eds.) FASE 2016. LNCS, vol. 9633, pp. 122–140. Springer, Heidelberg (2016). doi:10.1007/978-3-662-49665-7_8
12. Strüber, D., Schulz, S.: A tool environment for managing families of model transformation rules. In: Echahed, R., Minas, M. (eds.) ICGT 2016. LNCS, vol. 9761, pp. 89–101. Springer, Cham (2016). doi:10.1007/978-3-319-40530-8_6
13. Kappel, G., Langer, P., Retschitzegger, W., Schwinger, W., Wimmer, M.: Model transformation by-example: a survey of the first wave. In: Düsterhöft, A., Klettke, M., Schewe, K.-D. (eds.) Conceptual Modelling and Its Theoretical Foundations. LNCS, vol. 7260, pp. 197–215. Springer, Heidelberg (2012). doi:10.1007/978-3-642-28279-9_15
14. Acreţoaie, V., Störrle, H., Strüber, D.: VMTL: a language for end-user model transformation. Softw. Syst. Model. 1–29 (2016)

15. Panach, J.I., España, S., Moreno, A.M., Pastor, Ó.: Dealing with usability in model transformation technologies. In: Li, Q., Spaccapietra, S., Yu, E., Olivé, A. (eds.) ER 2008. LNCS, vol. 5231, pp. 498–511. Springer, Heidelberg (2008). doi:10.1007/978-3-540-87877-3_36

16. Ammar, L.B., Trabelsi, A., Mahfoudhi, A.: Incorporating usability requirements into model transformation technologies. Requir. Eng. **20**(4), 465–479 (2015)

17. Varró, D., Balogh, A.: The model transformation language of the VIATRA2 framework. Sci. Comput. Program. **68**(3), 214–234 (2007)

18. Geiß, R., Batz, G.V., Grund, D., Hack, S., Szalkowski, A.: GrGen: a fast SPO-based graph rewriting tool. In: Corradini, A., Ehrig, H., Montanari, U., Ribeiro, L., Rozenberg, G. (eds.) ICGT 2006. LNCS, vol. 4178, pp. 383–397. Springer, Heidelberg (2006). doi:10.1007/11841883_27

19. Schürr, A., Winter, A.J., Zündorf, A.: The PROGRES Approach: Language and Environment. Handbook of Graph Grammars and Computing by Graph Transformation. World Scientific Publishing Co. Inc., River Edge (1999)

20. Leblebici, E., Anjorin, A., Schürr, A.: Developing eMoflon with eMoflon. In: Ruscio, D., Varró, D. (eds.) ICMT 2014. LNCS, vol. 8568, pp. 138–145. Springer, Cham (2014). doi:10.1007/978-3-319-08789-4_10

21. Rensink, A., Schmidt, Á., Varró, D.: Model checking graph transformations: a comparison of two approaches. In: Ehrig, H., Engels, G., Parisi-Presicce, F., Rozenberg, G. (eds.) ICGT 2004. LNCS, vol. 3256, pp. 226–241. Springer, Heidelberg (2004). doi:10.1007/978-3-540-30203-2_17

22. Cabot, J., Clarisó, R., Guerra, E., De Lara, J.: Verification and validation of declarative model-to-model transformations through invariants. J. Syst. Softw. **83**(2), 283–302 (2010)

23. Strüber, D., Selter, M., Taentzer, G.: Tool support for clustering large meta-models. In: Workshop on Scalability in Model Driven Engineering (BigMDE), pp. 7:1–7:4 (2013)

24. Rentschler, A., Werle, D., Noorshams, Q., Happe, L., Reussner, R.H.: Remodularizing legacy model transformations with automatic clustering techniques. In: Workshop on Analysis of Model Transformations (AMT), pp. 4–13 (2014)

25. Baki, I., Sahraoui, H.: Multi-step learning and adaptive search for learning complex model transformations from examples. ACM Trans. Softw. Eng. Methodol. **25**(3), 20:1–20:37 (2016)

26. Buchmann, T., Westfechtel, B., Winetzhammer, S.: The added value of programmed graph transformations – a case study from software configuration management. In: Schürr, A., Varró, D., Varró, G. (eds.) AGTIVE 2011. LNCS, vol. 7233, pp. 198–209. Springer, Heidelberg (2012). doi:10.1007/978-3-642-34176-2_17

GRAPE – A Graph Rewriting and Persistence Engine

Jens H. Weber[(✉)]

LEADlab, Department of Computer Science,
University of Victoria, BC Victoria, Canada
jens@acm.org

Abstract. Graph-based data structures are fundamental to many applications in Computer Science and Software Engineering. Operations on graphs can be formalized as graph transformations or graph rewriting rules and a rich theoretical underpinning has been developed in the research community that supports reasoning about the properties of graph transformation systems. Various tools exist for developing graph transformations, including visual editors as well as textual languages that can be integrated with general purpose programming languages. This paper introduces *Grape* (Graph Rewriting and Persistence Engine), a hybrid, embedded Domain Specific Language (DSL) for Clojure. *Grape* is a lightweight approach to computing with persistent graphs within Clojure. It combines the ease of use of a textual DSL with a graphical visualization that is inlined with the program code when needed to aid comprehension and documentation of graph rewriting rules. Moreover, *Grape* supports persistence, programmed transactions and backtracking.

Keywords: Graph transformations · Tool support · DSL · Persistence · Clojure

1 Introduction

Graph-based data structures play an important role in many applications of Computer Science and Software Engineering. Operations on graphs can be formalized with graph transformation rules (also referred to as graph rewriting rules). A rich theoretical background exists on the formal properties of graph transformation systems (GTS) and various tools have been developed in support of their development [1, 2]. Current tool support ranges from visual development environments (with underlying transformation engines) to textual languages that may be integrated with general purpose programming languages. Visual development environments for GTS provide the benefit of a more intuitive, graphical way of specifying operations on graphs. However, these tools are often expensive to build and maintain. Moreover, visual development tools may pose usability challenges, as developers need to learn how to use them properly [3]. Other challenges pertain to the integration of visual programming in the overall software development lifecycle, such as the integration with other parts of a software program, configuration management and merging of different versions, etc.

Textual graph transformation languages provide a more lightweight approach to developing graph-based computations and avoid many of these challenges. However,

© Springer International Publishing AG 2017
J. de Lara and D. Plump (Eds.): ICGT 2017, LNCS 10373, pp. 209–220, 2017.
DOI: 10.1007/978-3-319-61470-0_13

textual graph rewriting rules may not be as easy to understand as their visual counterparts. Hybrid approaches in which graph transformations are specified textually but documented visually have been suggested as a compromise. However, textual programs and visual documentations are often not well integrated, which impede incremental and dynamic development of graph-based programs.

Another concern with existing graph transformation tools pertains to their scalability and the persistence of large graphs. Most current tools process graph models in main memory, which imposes practical limits to the scalability of their applications. Moreover, most current tools do not provide support for complex transactions of programmed graph rewrite operations, in the sense of the typical ACID properties (Atomicity, Consistency, Integrity and Durability).

This paper introduces *Grape* (Graph Rewriting and Persistence Engine) as a lightweight, hybrid GTS development tool that seeks to address the above concerns. *Grape* provides a lightweight, hybrid GTS extension to the Clojure programming language. Graph transformations are programmed textually, but visualized graphically, inline with the program code, within the LightTable general purpose text editor[1]. *Grape* utilizes the highly scalable Neo4 J graph database for graph persistence and transaction support. *Grape* programs provide full support for transactions, including backtracking. It has been made available under an open source license on Github.

The rest of this paper is structured as follows. The next section provides a short introduction to graph rewriting and graph transformation systems. Section 3 provides an overview of the work related to tool support for developing graph transformation systems in software applications. Section 4 provides an overview of the architecture of *Grape*, while Sect. 5 introduces the *Grape* DSL within Clojure and demonstrates the use of LightTable as a lightweight hybrid development environment. Finally, Sect. 6 provides concluding remarks and an outlook on future work.

2   Graph Rewriting and Graph Transformation Systems

2.1   Directed, Attributed, and Labeled (DAL) Graphs

A directed graph is a data structure that consists of a set of nodes N and a set of edges E, such that each edge $e \in E$ has a source $s(e)$ and target $t(e)$ in N. Labeled graphs allow nodes and edges to be labeled, i.e., a labeling function $l(o)$ associates each graph object $o \in N \cup E$ with a set of labels. Attributed graphs further allow the association of graph objects with attribute properties, i.e., an attribution function $a(o)$ associates each graph object with a set of key/value pairs.

2.2   Graph Transformation Rules

A graph transformation rule $L \rightarrow R$ consists of two graphs, commonly referred to as *left-hand side* (LHS) and *right-hand side* (RHS), respectively. The LHS specifies a

---

[1] http://lighttable.com.

subgraph pattern to find in a given graph (the *host graph*) and the RHS defines how a found subgraph is to be rewritten as a result of the transformation. In other words, LHS acts as a precondition of the rule, while RHS specifies its post-condition. Graph transformation rules are applied in a three-stepped process:

1. **Match:** The hostgraph is searched for a subgraph that matches the rule's LHS.
2. **Delete:** Graph objects in the rule's LHS that are not in the rule's RHS are deleted from the hostgraph.
3. **Add:** Graph elements in the rule's RHS that are not included in its LHS are added to the host graph.

When a rule matches multiple subgraphs in the host graph, a match is chosen non-deterministically.

The above process of applying graph transformation rules may result in a structure that is not a graph. This is the case when a node is deleted that is a source or a target of an edge in the host graph that is not part of the match found for the rule's LHS. The graph transformation community has developed different approaches on how to prevent this situation. One approach is to permit the deletion of such a node and to delete all edges that may be left "dangling" in the host graph. This approach is based on the "*single pushout*" (SPO) theory for graph transformations [1]. Another approach is to prohibit the application of a transformation rule in cases where its execution would result in dangling edges. This more restrictive condition is commonly referred to as the *gluing condition* in the *double-pushout* (DPO) theory for graph transformations [1].

Different approaches exist as well with respect to the type of morphism that is used to find a match for a rule's LHS in the host graph. Isomorphic matching requires that each object in a rule's LHS matches a distinct graph object in the host graph, while homomorphic matching allows different objects in a rule's LHS to match to the same graph object in the host graph.

Graph transformation rules may also have a set of negative application conditions (NACs) that may be used to prevent rule application in certain contexts. NACs are an important concept for many practical applications of GTS [4]. NACs can be specified as graph patterns (and conditions on attributes) that extend a rule's LHS. The application of graph transformation rules with NACs becomes a four-step process:

1. **Match:** The hostgraph is searched for a subgraph that matches the rule's LHS.
2. **Check:** Attempt to extend the matched subgraph with a match for any of the rule's NACs. If this is possible, prevent rule application in this context.
3. **Delete:** Graph objects in the rule's LHS that are not in the rule's RHS are deleted from the hostgraph. (Validate gluing condition for DPO rewriting approach.)
4. **Add:** Graph elements in the rule's RHS that are not included in its LHS are added to the host graph.

A graph transformation system (GTS) is defined as a set of graph transformation rules. A graph grammar is a GTS with a defined start graph. Graph grammars are commonly used for defining and parsing graph-based languages. In this paper, we are less concerned with the definition of graph-based languages and rather focus on engineering applications of graph rewriting. Such applications typically require imperative control structures to govern the execution of graph transformation rules. A programmed GTS is a GTS that has been associated with an imperative control program.

3 Related Work: Tool Support for Graph Rewriting

3.1 Visual Tools

PROGRES is an integrated development environment for visual developing of programmed GTS [5]. PROGRES provides a powerful specification language and uses a graph database for scalability and persistence with support for complex transactions and backtracking. However, PROGRES lacks integration with general purpose programming languages and its development has been discontinued.

FUJABA supports the development of programmed GTS for Java [6]. Control structures are specified using "story diagrams", a combination of activity diagrams and graph rewriting rules. Graph transformations are carried out in main memory. Graphs can be serialized for file-based storage. FUJABA generates Java code, which can be integrated with general purpose Java programs. FUJABA does not support transactions and backtracking.

AGG is a visual development environment for GTS [7, 8]. Transformation rules are executed based on an interpreter and the graph is held in main memory. The definition of control structures is supported. An API allows the integration with the Java general purpose programming language.

GROOVE is another visual GTS development environment that is particularly suitable for formal verification and state space exploration [9]. The graph is held in main memory and transformations are executed by an interpreter. Control structures are provided in form of a dedicated scripting language. Backtracking and transactions are not supported. Integration with general purpose programming languages is possible through an internal (undocumented) API.

Henshin is a visual graph transformation tool based on the Eclipse modelling framework (EMF) [10]. Graph transformations are carried out by an interpreter that can be interfaced with general purpose programming languages (Java) through an API. Graphs are kept in main memory. No transactions or backtracking is supported.

3.2 Textual Tools

Viatra is an Eclipse plugin that provides a textual language for specifying graph transformations [11, 12]. Control structures are specified using abstract state machines.

GrGen provides a textual language to define graph transformations on object graphs held in main memory (C# or Java) or kept in a relational database [13]. GrGen generates C# code or .net assemblies.

SDMlib is an internal DSL for graph transformations with Java [14]. Graphs are kept in main memory and can be persisted in a file. Graphs and graph transformations can be visually documented. Transactions and backtracking is not supported.

FunnyQT is an internal graph transformation DSL for Clojure [15]. Homomorphic as well as isomorphic graph pattern matching is supported. FunnyQT provides a framework for in-place graph rewriting with arbitrary Clojure actions. The semantics of rewriting rules is not grounded in a particular theory (such as DPO or SPO). Graphs are held in main memory and transactions are not supported.

4 The Grape Architecture

Grape provides an internal domain-specific language (DSL) for programming GTS with Clojure (cf. Fig. 1). *Grape* uses the Neo4J graph database for storing the host-graph, i.e., the graph is not held in main memory. Graph transformation rules defined in the *Grape* DSL are translated to Cypher, Neo4J's native query language. Cypher provides powerful constructs for graph pattern matching, which are leveraged by *Grape*. *Grape* also provides a visualizer for graph transformation rules based on GraphViz [16]. *Grape* does not depend on any particular development tool or IDE, but provides a convenient integration with LightTable, which allows developers to visualize their graph transformation rules "in line" with their textual definition. Independently of the editor used, the *Grape* visualizer also provides functions to generate visual representations of GTS in the file system. Neo4J also provides an extensible graph browser that can be used to visualize graphs.

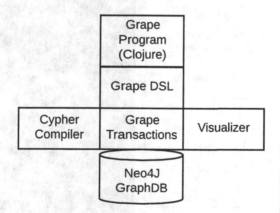

Fig. 1. The *Grape* architecture

Grape provides support for complex transactions of programmed graph transformations with full support for backtracking. This functionality is based on Neo4J's flat transaction model and implemented in *Grape's* transaction module. (Neo4J native "flat" transaction model needs to be extended to support nested transactions, as required for the desired backtracking behaviour.)

5 A Taste of Grape – Introduction to Programming with Grape

5.1 Simple Rule Definition and Execution

The *Grape* DSL uses native Clojure syntax. Graphs are schema-less (untyped), following the schema-less design philosophy of no-SQL databases such as Neo4 J. Figure 2 provides a first example of the *Grape* language. It defines a new GTS (using the `gts` form) and a simple transformation rule that matches two graph nodes that are connected with a `works_for` edge in order to replace that edge with a new `Contract` node with `employee` and `employer` edges.

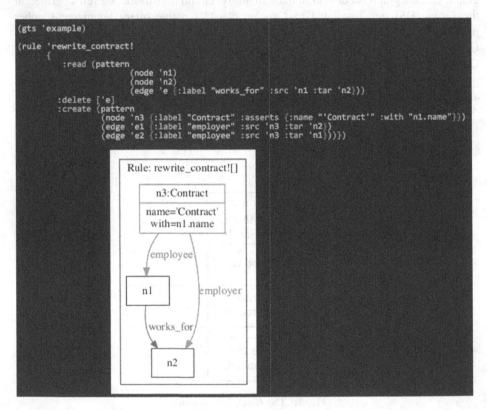

Fig. 2. Definition of a GTS and a simple transformation rule

As the example in Fig. 2 shows, *Grape* rules have three main parts: *read*, *delete* and *create*. (Of course, some of these parts may be missing. For example, a rule that simply creates a graph structure will not have a *read* or *delete* part.) Nodes and edges are identified by id symbols, e.g., 'n1. Labels are defined using the `:label` key and assertions on attributes are defined using the `:asserts` key. In the above example, two attributes are defined for the new `Contract` node: the first attribute (name) is assigned the literal value "Contract", while the second attribute (`with`) is assigned the value of the `name` attribute of node n1.

The visualization of the rule in Fig. 2 is automatically generated by *Grape* and inlined after the textual definition (if LightTable is used as the editor). The visualization uses the popular representation of graph transformation rules where LHS and RHS are merged into the same graph, with red colours showing deleted graph elements and green colour showing new ones. *Grape* programmers who do not use LightTable can still make use of the rule visualization by generating visual documentation in the project's file system.

Once a rule has been defined (as above), it can be invoked simply by calling an equally named function, i.e., by calling (`rewrite_contract!`) in the above example. Calling this function will return true if the `rule` could be applied and `false` otherwise. Its invocation will non-deterministically select a possible match and attempt its transformation. The usual Clojure control structures can be used with this function. For example, if all `works_for` occurrences are to be rewritten, a programmer may simply use (`while` (`rewrite-contract!`)).

Grape does not implement its own tool for visualizing the state of the hostgraph, since Neo4J provides a powerful graph browser as part of its community edition. Figure 3 shows a hostgraph visualized with the Neo4J browser for the Ferryman example discussed at the end of this section.

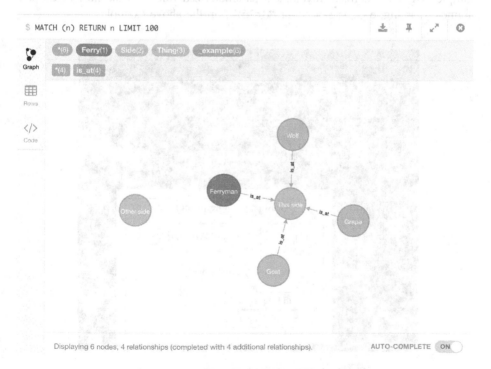

Fig. 3. Hostgraph visualization using Neo4J's graph browser

5.2   Customizing Matching and Rewrite Semantics

Grape performs isomorphic graph matching by default. This means that in the above example, self-employment relationships (where n1 and n2 match the same node in the host graph) would not be matched. If homomorphic matching is desired, a :homo key can be added to the definition of the *read* pattern.

Furthermore, *Grape* rules implement SPO rewrite semantics by default, i.e., any "dangling" edges that may arise from deleting nodes are automatically deleted as well. If the more restrictive DPO semantics is desired, an :dpo key can be added to a rule, which results in checking the *gluing condition* prior to executing the transformation.

5.3   Parameters and NACs

Grape transformation rules can by parameterized and contain an arbitrary number of NACs. Figure 4 provides an example for a rule "promote!" with a formal parameter and one NAC. It searches for a Worker who works_for an Employer with a given name (parameter) and replaces the Worker node with a Director node, if that worker does not also work for another Employer (i.e., if there is not a work_for edge from node w to another node in the host graph). Graph patterns defined in NACs are visualized with dashed lines and using a different colour for each defined NAC, if multiple NACs are defined. Invoking a parameterized rule uses the normal Clojure parameter passing, e.g., (promote! "John") in the above example.

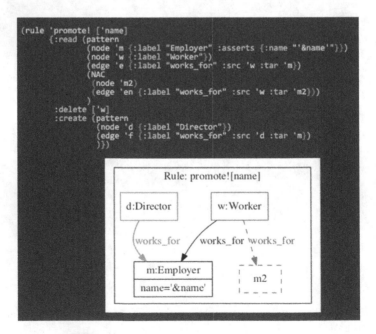

Fig. 4. A parameterized rule with a single NAC

5.4 Transactions

Grape supports transactions of complex graph rewriting operations consisting of multiple transformation rules, with full backtracking support. Transactions are created using the `transact` form. The left-hand side of Fig. 5 shows a simple transaction consisting of a sequence of programmed transformation rules.

```
(transact                          (transact
   (apl 'rule1)                       (apl 'rule1 p)
   (apl 'rule2)                       (bind 'out 'n)
   ..                                 (apl 'rule2 (consult 'out))
   (apl 'ruleN))                      ..)
```

Fig. 5. Defining simple transactions (left) and defining parameterized transaction operations with parameter passing (right).

Note that *programmed* transformation rules require the `apl` form to define each rule application. Writing `(rule1)` instead of `(apl 'rule1)` would cause the Clojure REPL to execute "rule1" at the time of defining the transaction. Of course, we could have used a macro instead of a regular Clojure function for defining the transact form to prevent this behaviour. However, we intentionally decided to avoid macros in the development of *Grape* to keep the code simple and functional.

Grape also allows parameter passing between different transformation rules in a transaction. This is realized using two forms: `bind` and `consult`. The first form binds a graph element (node or edge) matched by the previously executed graph transformation rule to a symbol, while the second form (`consult`) dereferences the bound graph object for the purpose of passing it to a subsequent transformation rule. The right-hand side of Fig. 5 shows an example. Here, the graph element n matched in `rule1` is bound to symbol `out` and then passed to `rule2` as a parameter.

```
(defn tx [p]
   (transact                         (attempt (tx! "hello"))
      (apl 'rule1 p)
      ..
   ))
```

Fig. 6. Defining and attempting to execute a parameterized transaction.

The `transact` form returns a value that can be passed to the `attempt` form for execution. Of course, parameterized transactions can be defined as regular Clojure functions, using the `defn` form. Figure 6 shows an example of defining a transaction with an example parameter p (left) and attempting to execute it with an example argument "hello" (right). The result of executing an `attempt` form is *true* or *false*, depending on whether the transaction succeeded.

5.5 Control Structures

As *Grape* is an embedded DSL the full breadth of control structures of the host language (Clojure) can be utilized for programming with graph transformation rules. (An example was given at the end of Sect. 5.1.) Moreover, since Clojure is a JVM-based language, other JVM languages can also be used.

A limitation of using the GPL host language's control structures is that they provide no support for backtracking in operations that compose multiple graph transformations. *Grape* provides a set of native DSL control structures that can be used to program complex graph operations where backtracking is desired. In particular, *Grape* provides control structures for loops, non-deterministic choice, and negative graph tests.

Figure 7 shows an example program for solving the well-known Ferry Crossing puzzle [17]. In that problem, a ferryman has the task to safely transport three items across the river. He can only take one item at a time. Two unsafe states exist: (1) the wolf will eat the goat and (2) the goat will eat the cabbage, if left unsupervised.

```
(until 'all_on_the_other_side?
       (transact (choice (apl 'ferry_one_over!)
                         (apl 'cross_empty!))
                 (avoid (apl 'wolf-can-eat-goat?)
                        (apl 'goat-can-eat-grape?)))))
```

Fig. 7. Example using *Grape* control structures

The until form is used to define a loop with a break condition given as its first argument, followed by a set of *Grape* transactions that are to be executed in each iteration. The break condition 'all_on_the_other_side? is a *graph test*. It is defined as a regular *Grape* graph transformation rule that only has a *read* part. The choice form takes a list of *Grape* rule applications or transactions and non-deterministically selects one of them for execution. Finally, the avoid form takes a list of graph tests and checks whether any of them have a match in the host graph. In that case, the current state of the graph exploration is considered a failure and the program will backtrack. Note that the avoid form is not strictly necessary for expressiveness, since *Grape* rules support the definition of NACs. However, we believe that its existence may increase the readability and conciseness of programs.

The program in Fig. 6 is an example for implementing a graph exploration search algorithm in *Grape*. Essentially, the program implements a forward rule-chaining algorithm. Rule selection by the choice operator is non-deterministic, which means that the above search is not guaranteed to find a solution (and to terminate). Of course, the graph transformation rules can be defined to limit the search space. A common approach is to define a cost for each ferry crossing and allocate a budget. This can be done using graph attribute assignments and application conditions. In the future, we are interested in extending *Grape* with a heuristics guided choice operator that uses a utility function to aid the prioritization of alternative rule applications.

6 Conclusions and Future Work

The choice of a graph transformation tool ultimately depends on the application it is used for. Several graph transformation tools have been developed and made available. We found that for our applications, graph persistence, scalability, transaction handling and the ability to easily integrate with general purpose programming languages were important requirements. Moreover, we found that the user interfaces of fully visual graph transformation tools often create usability challenges and do not integrate seamlessly with modern, distributed software development practices, e.g., versioning, merging, test-driven development, etc.

In this paper, we have introduced *Grape* as a lightweight, hybrid graph transformation engine embedded in Clojure. *Grape* is highly scalable, as graphs are not kept in main memory but in a graph database (Neo4J scales to graphs consisting of tens of billions of nodes). *Grape* is considered a "hybrid" tool, since rules are authored textually but visualized graphically "in line" with the textual code (if LightTable is used as the editor). Since *Grape* programs are authored with an embedded DSL in Clojure, they can easily be interfaced with the rest of a software program. Moreover, *Grape* provides full support for complex transactions, including backtracking.

Grape has been made available for public use on Github[2]. So far, we have used *Grape* in the development of one small-sized application in the medical domain. The source code for this example application is also available on Github[3]. While in this first application we did not make use of some of *Grape*'s advanced concepts (such as complex transactions and backtracking), it illustrates nicely how easy it is to integrate a *Grape* GTS with the rest of a typical Web-based software system.

There are many avenues for future work on improving *Grape*. We already mentioned at the end of the last section the plan to add a heuristics guided choice operator, to direct the selection of alternative graph transformation rules in complex transactions. Moreover, *Grape* currently operates on untyped graphs. This provides a great degree of flexibility but also increases the likelihood of specification errors. We will be adding the option of working with typed graphs in the future. Another worthwhile extension to *Grape* would be the addition of path expressions. Neo4J provides support for powerful graph expressions in its query language Cypher. We expect to be able to use this feature as a basis for implementing path expressions in *Grape*.

Finally, we are intending to extend *Grape* with respect to supporting bidirectional transformations between graph structures. Triple Graph Grammars (TGG) have been proposed and successfully used for bidirectional graph model synchronization problems. Their integration in *Grape* is planned for a future release [18].

---

[2] https://github.com/jenshweber/grape.

[3] https://github.com/sdiemert/app-project.

References

1. Rozenberg, G.: Handbook of Graph Grammars and Computing by Graph Transformation - Volume 1: Foundations. World Scientific Publishing Company (1999)
2. Ehrig, H., Engels, G., Kreowski, H.-J., Rozenberg, G.: Handbook of Graph Grammars and Computing by Graph Transformation - Volume 2: Applications, Languages, Tools. World Scientific Publishing Company (1999)
3. Rouly, J.M., Orbeck, J.D., Syriani, E.: Usability and suitability survey of features in visual ides for non-programmers. In: Proceeding of the 5th Workshop on Evaluation and Usability of Programming Languages and Tools, New York, NY, USA, pp. 31–42 (2014)
4. Habel, A., Heckel, R., Taentzer, G.: Graph grammars with negative application conditions. Fundam. Informaticae **26**(3), 287–313 (1996)
5. Schürr, A., Winter, A.J., Zündorf, A.: Graph grammar engineering with PROGRES. In: Schäfer, W., Botella, P. (eds.) ESEC 1995. LNCS, vol. 989, pp. 219–234. Springer, Heidelberg (1995). doi:10.1007/3-540-60406-5_17
6. Nickel, U., Niere, J., Zündorf, A.: The FUJABA environment. In: Proceeding of the 22nd Intl Conference on Software Engineering, New York, NY, USA, pp. 742–745 (2000)
7. Taentzer, G.: AGG: a graph transformation environment for modeling and validation of software. In: Applications of Graph Transformations with Industrial Relevance, pp. 446–453 (2003)
8. Runge, O., Ermel, C., Taentzer, G.: AGG 2.0 – new features for specifying and analyzing algebraic graph transformations. In: Applications of Graph Transformations with Industrial Relevance, pp. 81–88 (2011)
9. Ghamarian, A.H., de Mol, M., Rensink, A., Zambon, E., Zimakova, M.: Modelling and analysis using GROOVE. Int. J. Softw. Tools Technol. Transf. **14**(1), 15–40 (2012)
10. Arendt, T., Biermann, E., Jurack, S., Krause, C., Taentzer, G.: Henshin: advanced concepts and tools for in-place EMF model transformations. In: Model Driven Engineering Languages and Systems, pp. 121–135 (2010)
11. Balogh, A., Varró, D.: Advanced model transformation language constructs in the VIATRA2 framework. In: Proceedings of the 2006 ACM Symposium on Applied Computing, New York, NY, USA, pp. 1280–1287 (2006)
12. Bergmann, G., et al.: Viatra 3: a reactive model transformation platform. In: Theory and Practice of Model Transformations, pp. 101–110 (2015)
13. Geiß, R., Batz, G.V., Grund, D., Hack, S., Szalkowski, A.: GrGen: a fast SPO-based graph rewriting tool. In: Graph Transformations, pp. 383–397 (2006)
14. Priemer, D., George, T., Hahn, M., Raesch, L., Zündorf, A.: Using graph transformation for puzzle game level generation and validation. In: Graph Transformation, pp. 223–235 (2016)
15. Horn, T.: Graph pattern matching as an embedded clojure DSL. In: Graph Transformation, pp. 189–204 (2015)
16. Ellson, J., Gansner, E., Koutsofios, L., North, S.C., Woodhull, G.: Graphviz— open source graph drawing tools. In: Graph Drawing, pp. 483–484 (2001)
17. Gasarch, W.: Review of algorithmic puzzles by Anany Levitin and Maria Levitin. SIGACT News **44**(4), 47–48 (2013)
18. Schürr, A., Klar, F.: 15 years of triple graph grammars. In: Graph Transformations, pp. 411–425 (2008)

Table Graphs

Albert Zündorf[1]([⊠]), Daniel Gebauer[2], and Clemens Reichmann[2]

[1] Kassel University, Kassel, Germany
zuendorf@uni-kassel.de
[2] Vector Informatik GmbH, Stuttgart, Germany
{Daniel.Gebauer,Clemens.Reichmann}@vector.com

Abstract. Inspired by the PREEvision tool of Vector Informatik GmbH Stuttgart, we have extended SDMLib with so-called Table Graphs. Table Graphs model the matches of graph rewrite rules as explicit graphs added to a given host graph. Table graphs allow to do relational operations like filter and projection. We also provide support for the extension of matches with new nodes or attribute values. Table Graphs also allow to do many spreadsheet like computations via simple graph rewrite rules. Finally, Table Graphs may be exported as HTML Tables or CSV files. This allows e.g. to generate a nice PIE chart from a graph query result.

Keywords: Graph grammars · Graph transformations · Fujaba · SDMLib

1 Introduction

The PREEvision tool of Vector Informatik GmbH, Stuttgart, [1] supports the design and modeling of the complete car electronic including ECUs, sensors, actors, wiring harness, software components, communication bus system, and signals. The PREEvision tool is graph based and utilizes a complete graph transformation engine developed by Clemens Reichmann. This graph engine realizes complex editing and analysis operations and e.g. model to model transformations between different levels of abstraction. Still, the users of the PREEvision tool asked for better support of frequent statistic queries like "list of all components (within the motor area)", "total costs", "total weight", "total length of all wires", etc. Therefore, in their Phd thesis [2] Daniel Gebauer and Johannes Matheis [3] extended the PREEvision graph engine with *Table Graphs*. At first, a Table Graph lists all matches of a graph rewrite rule. This may be used e.g. to count all components of a certain type. In addition, Table Graphs provide filter operations that may narrow down the interesting matches. Additional operations allow to extend a Table Graph with additional columns e.g. for attribute values like sizes, weights, or prices. Then you may compute column sums, minimum, maximum, or average, etc. You may also forward a table graph e.g. to a spreadsheet program. The spreadsheet program may than produce colorful charts for your next PowerPoint presentation.

© Springer International Publishing AG 2017
J. de Lara and D. Plump (Eds.): ICGT 2017, LNCS 10373, pp. 221–230, 2017.
DOI: 10.1007/978-3-319-61470-0_14

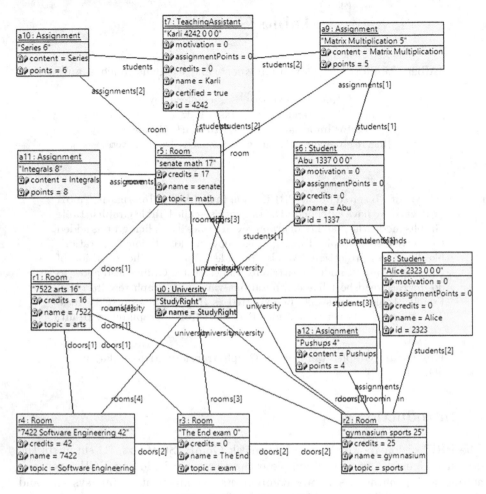

Fig. 1. Example graph

I (Albert) was the second reviewer of the PhD thesis of Daniel Gebauer. Impressed by the possibilities provided by Table Graphs, I decided to add similar features to the SDMLib graph engine [4,5]. In SDMLib, Table Graphs are first class citizens, i.e. Table Graphs become actually part of the original host graph. Thus, Table Graphs may be further analyzed and extended with the help of other graph transformations. Similarly, a set of predefined graph transformations may be used to filter Table Graphs or to extend a Table Graph with additional columns or to derive new Table Graphs from existing ones or to do some attribute computations. In addition, Table Graphs may be exported as HTML tables or as CSV files for use in web pages or in spreadsheet programs, respectively.

This paper outlines the graph structure of Table Graphs, how to specify typical graph transformations on Table Graphs and some predefined graph transformations frequently used on Table Graphs and some export functionality.

2   Background: SDMLib Graph Transformation Rules

As one of the reviewers pointed out, we need to introduce SDMLib graph transformation rules in order to discuss Table Graphs. Thus, this section gives a short introduction into SDMLib.

Figure 1 shows a graph of a university with some rooms and some students and some assignments. On this graph we may run the simple graph query shown in Fig. 2. This graph query matches all pairs of University and Room nodes connected via a rooms edge. Actually, SDMLib generates a model specific Java API for the creation and execution of graph rewrite rules. Thus the graph rewrite rule shown in Fig. 2 is created by the following Java code:

```
1  public Table findRooms(University u) {
2      UniversityPO u1 = new UniversityPO(u);
3      RoomPO r2 = u1.createRooms();
4      return u1.createResultTable();
5  }
```

Listing 1.1. Example Graph Query in Java

Line 2 creates the `UniversityPO` object u1 that is a so-called *Pattern Object* that matches host graph objects of type `University`. In addition, the host graph object passed as parameter u is provided to the pattern object a match candidate. Thus, u1 will match u. We call the pattern object u1 *bound*, cf. the corresponding stereotype in the graphical representation of our graph rewrite rule in Fig. 2. Line 3 of Listing 1.1 extends the graph rewrite rule with a `rooms` link connecting u1 to a new `RoomPO` pattern object r2. Finally, line 4 of Listing 1.1 calls method `createResultTable()` on pattern object u1. (Any pattern node may serve as pars-pro-toto for the whole pattern.) Method `createTable()` is discussed in the next section. Once the graph transformation rule is constructed, it may be rendered as an SVG graphics e.g. with the command u1.`dumpDiagram("diag1")`.

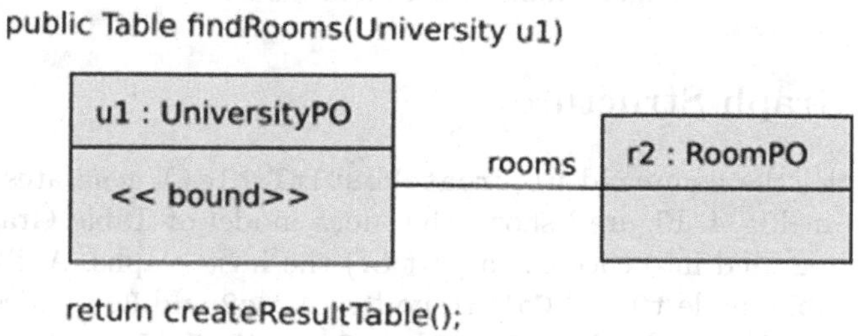

public Table findRooms(University u1)

```
┌─────────────────────┐                    ┌─────────────────────┐
│  u1 : UniversityPO  │       rooms        │   r2 : RoomPO       │
│                     │────────────────────│                     │
│   << bound>>        │                    │                     │
└─────────────────────┘                    └─────────────────────┘
```

return createResultTable();

Fig. 2. Example graph query

The operations used to create pattern elements may have additional parameters allowing to mark the pattern elements with <<create>> or <<destroy>> stereotypes, cf. Figs. 5 and 6. In the SVG graphics such elements are rendered with green or red color, respectively. (In an earlier version of this paper, we used a somewhat inconsistent color coding of pattern elements. We have addressed the corresponding reviewer comments and now adapt the color coding proposed by Henshin [6].)

In addition, SDMLib allows to group pattern elements into (nested) sub-patterns. In our graphical representation sub-patters are shown by extra rectangles, cf. Fig. 5. Sub-patterns may have additional stereotypes marking them as <<optional>> or <<allmatches>> or <<NAC>>. Optional sub-patterns do not affect the applicability of a graph rewrite rule but allow to handle additional host graph elements in the area of a core match, if such elements exist. Optional sub-pattern are rendered as dashed black rectangles with the stereotype <<optional>>. Allmatches sub-pattern are iterated or amalgamated, i.e. they match as often as possible. Allmatches sub-pattern are rendered as two stacked solid black rectangles with the stereotype <<allmatches>>. NAC sub-pattern represent negative application conditions. NAC sub-pattern are rendered as a blue solid rectangle with a blue forbidden icon in the top left corner, cf. Fig. 5.

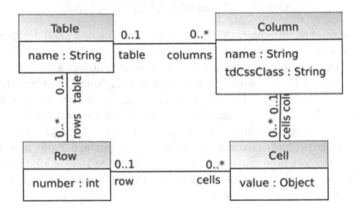

Fig. 3. Table graph meta model

3   Table Graph Structure

In our example, the command u1.createResultTable() generates the Table Graph shown in Fig. 4. Figure 3 shows the meta model of Table Graphs. Table Graphs are integrated into (become a part of) the host graphs. A Table Graph consists of a Table node t0 and Column nodes c1, c2 and Row nodes r3 - r7. The table content is modeled by Cell nodes c8 - c17. Cell nodes may contain attribute values or refer to host graph nodes via value edges, cf. nodes c16, c17 and u19, r18, respectively. (Other host graph nodes are omitted for brevity.) In Fig. 4 the cell nodes show a String derived by the toString() method of the corresponding host graph node.

Each Row node in our Table Graph represents one match of our example graph query. Each Cell node represents the match of one node from our graph

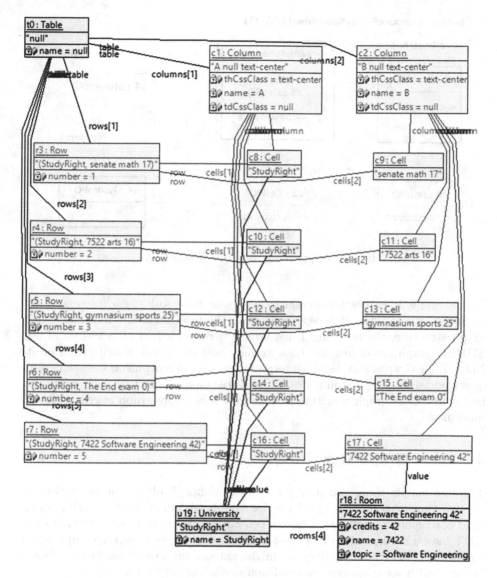

Fig. 4. Example table graph (cutout)

query to a node in our host graph. All `Cell` nodes belonging to the same graph query node are attached to a `Column` node via a `column--cells` link. Ideally, the `Column` nodes would refer to the corresponding graph query node, unfortunately, the graph query nodes are not yet first class citizens of the SDMLib graphs. Thus, the `Column` nodes are just marked by running letters `A, B, ....` Similarly, the `Row` nodes have running numbers `1, 2, ....` This schema has been borrowed from spreadsheet programs, intentionally. However, this schema also resembles the structure of relational database tables.

public int removeEmptyRoomRows(Table t1)

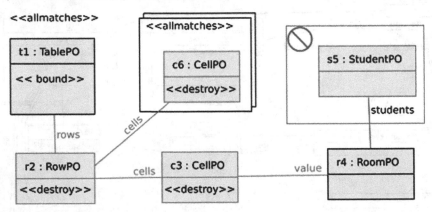

Fig. 5. Remove empty room rows rule (Color figure online)

Generally, it is pretty common to represent the results of a query as a set of tuples or as a table. However, by turning the table structure into an explicit graph structure, the resulting Table Graph becomes a first class citizen of our SDMLib graph structures and thus we may now run graph rewrite rules on it. Note, Table Graphs may also be interpreted as a set of morphisms between graph query nodes and host graph nodes. Thus, Table Graphs also turn such morphism into a first class graph structure and we may now apply graph rewrite rules on morphisms.

4   Table Graph Rewriting

The first simple thing you may want to do with a Table Graph is to filter or remove some matches. The graph rewrite rule shown in Fig. 5 removes all rows r2 (and corresponding cells) that have a cell c3 that refers to a room r4 that does NOT have a student s5 in it. (The forbidden sign marks a negative application condition.) (Note, alternatively we might extend our findRooms() rule shown in Fig. 2 with some application condition excluding empty rooms.)

Similarly, one may use a graph rewrite rule to drop a column (e.g. with a certain column name). However, our Graph Table class provides a predefined method withoutColumn(String... colName) that does this job.

More frequently we use rules that extend a table with new columns. The rule addTopicColumn() shown in Fig. 6 first creates a column node with name "Topic". In addition it has an iterated sub-pattern that matches all rows r3 of table t1 and cells c4 with a value link to some room r5. For each sub-match, a cell c6 with value r5.topic is created and attached to its column and row.

As adding columns with attribute values is a quite frequent task, our Graph Table class provides method createColumns(String colName, RowLambda l)

that takes a column name and an Java 8 lambda expression as parameter, cf.
Listing 1.2.

```
1   table.createColumns("Credits",
2   row -> ((Room) row.getCellValue("B")).getCredits());
3   table.createColumns("Students",
4   row -> ((Room) row.getCellValue("B")).getStudents().size());
5   table.withoutColumns("A", "B");
```
Listing 1.2. Add Attribute Columns

After adding a "Topic", a "Credits", and a "Students" column and drop-
ping the original columns "A" and "B", we may dump our table e.g. as HTML
table and open it within a browser, cf. Fig. 7. We may also dump the table in CSV
format and import it into a spreadsheet program and e.g. produce a nice pie chart
from the credits column, cf. Fig. 8. Alternatively, we may compute charts directly.
Our Table Graphs also support some spreadsheet functionality, directly, e.g. the
path expression `table.getColumn("Credits").getValueSum()` computes the
sum of all credits for our example.

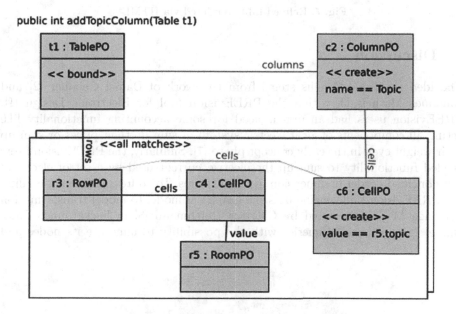

Fig. 6. Add Topic Column Rule (Color figure online)

With respect to relational table operations, we might also model a graph
rewrite rule doing a (natural) join of two tables. We leave this as an exercise for
the interested reader.

Topic	Credits	Students
math	17	1
arts	16	0
sports	25	2
exam	0	0
Software Engineering	42	0

Fig. 7. Refined table rendered via HTML

5   Discussion

The idea for Table Graphs stems from the work of Daniel Gebauer [2] and Johannes Matheis [3] within the PREEvision tool for Electronic Design [1]. PREEvision users had an urgent need for some accounting functionality like listing all components of a car electric system or sum up their prices or sum up their weight even in the early concept phase [7]. Similarly, the PREEvision users needed functionality to sum up the electric current used by a set of electrical components in order to dimension fuses and wires. To address such functionality, the PREEvision table engine uses the graphical model to model transformation language M^2ToS presented by Clemens Reichmann [8] in background. First, Daniel extended model queries with the possibility to annotate its nodes and

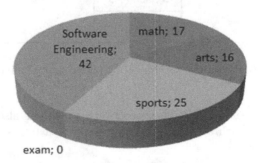

Fig. 8. Pie Chart for our example.

attributes to become part of the resulting Table Graph. Then, he added a data flow oriented graphical notation for filter, projection and aggregation operations. Similarly, Daniel allows subsequent graph rewrite rules (based on M^2ToS [8]) to operate on intermediate Table Graphs. This Table Graph functionality was very well accepted by the PREEvision users as it enables them to do frequent jobs that formerly required some tedious manual programming for data export and then import into a spreadsheet program and then doing the accounting.

Generally, the idea of reporting the results of a graph query is quite obvious. In Fujaba and SDMLib we usually did this by creating some kind of marker nodes within a graph rewrite rule and by connecting the marker node with the matched host graph node via some edges. Such marker nodes somehow correspond to the Row nodes of our Table Graphs. Sometimes you have multiple host graph nodes of the same type within one match. In such cases you want to mark the different roles of these host graph nodes e.g. by using different edge labels. As Fujaba and SDMLib are strongly typed, you cannot introduce new edge labels on the fly but you need to extend the graph schema first. To avoid such efforts, we frequently used attributed edges which means in our case an edge - node - edge combination where the middle node carries e.g. a key attribute. In Table Graphs this would correspond to Cell nodes with a key attribute holding e.g. the column name. While such marker nodes with attributed edges work perfectly well, they do not explicitly model that you have a set of matches that share a similar structure, i.e. that all employ the same set of attributed edges and keys. Within Table Graphs the Column nodes serve exactly this purpose. Table Graph Column nodes facilitate a number of column related operations like drop column or add column. Having an explicit Column node especially facilitates the management of Table Graphs via graph queries and graph rules.

Formerly, SDMLib did not create the result of a graph query as an explicit data structure. Instead, we used Java control structures, for example while (rule.hasMatch()) loops where in each iteration one asked the elements of the graph query for their current match. While this works in principle we somehow had the same problems as the PREEvision users that certain accounting tasks were tedious and were done in plain Java. The real contribution of Table Graphs is to make the graph query results an explicit graph structure that is open for the application of subsequent graph rewrite rules.

Within the PREEvision tool, Daniel finally dropped explicit Cell nodes and connected the corresponding host graph nodes to their rows and columns, directly. The reason is the huge number of Cell nodes created by matches. You easily have multiple times the number of Cell nodes compared to the number of host graph nodes. In SDMLib we did not yet adopt this idea in order to maintain the Table Graph structure more expli and more direct.

6   Summary

We have introduced Table Graphs as an explicit graph representation of graph query results. Table Graphs make graph query and graph rewrite rule matches an explicit data structure that may be used as input for further graph transformations. Table Graphs also resemble the structure of relation database tables and the structure used in spreadsheet programs. Thus, to some extend Table Graphs are a bridge between graph technology and table technologies and theories. While this is not rocket science, Table Graphs turned out to be very handy for quite a number of cases. As it is very simple to add Table Graph concepts to any graph engine, we propose that any graph engine should provide Table Graph functionality as it serves very well for many accounting functionalities.

References

1. Muller-Glaser, K.D., Reichmann, C., Graf, P., Kuhl, M., Ritter, K.: Heterogeneous modeling for automotive electronic control units using a case-tool integration platform. In: IEEE International Symposium on Computer Aided Control Systems Design, pp. 83–88. IEEE (2004)
2. Gebauer,D.J.: Ein modellbasiertes, graphisch notiertes, integriertes Verfahren zur Bewertung und zum Vergleich von Elektrik/Elektronik-Architekturen. PhD thesis, Dissertation, Karlsruhe, Karlsruher Institut für Technologie (KIT) (2016)
3. Matheis, J.: Abstraktionsebenenbergreifende Darstellung von Elektrik/Elektronik-Architekturen in Kraftfahrzeugen zur Ableitung von Sicherheitszielen nach ISO 26262. PhD thesis, Dissertation, Karlsruhe, Karlsruher Institut für Technologie (KIT) (2009)
4. Norbisrath, U., Jubeh, R., Zündorf, A.: Story Driven Modeling. CreateSpace Publishing Platform (2013)
5. Story Driven Modeling Library (2014). http://sdmlib.org/
6. Henshin (2017). https://www.eclipse.org/henshin/
7. Matheis, J., Gebauer, D., Kuhl, M., Reichmann, C., Muller-Glaser, K.D.: Vorstellung einer methodik zur e/e-architekturmodellierung und -bewertung in der fruhen konzepthase. In: 26. Tagung 'Elektronik im Kraftfahrzeug' vom Haus der Technik Essen e.V., Dresden, Germany (2006). Haus der Technik
8. Reichmann, C.: Grafisch notierte Modell-zu-Modell- Transformationen fr den Entwurf eingebetteter elektronischer Systeme. PhD thesis, Dissertation, Karlsruhe, Universität Karlsruhe (TH), 2005 (2005)

Author Index

Printed in the United States
By Bookmasters